The Limits of Technology and the End of History

Yefim Kats

The Limits of Technology and the End of History

Marx and Beyond

Yefim Kats
Plymouth Meeting, PA, USA

ISBN 978-3-031-69544-5   ISBN 978-3-031-69545-2 (eBook)
https://doi.org/10.1007/978-3-031-69545-2

© The Editor(s) (if applicable) and The Author(s), under exclusive license to Springer Nature Switzerland AG 2024

This work is subject to copyright. All rights are solely and exclusively licensed by the Publisher, whether the whole or part of the material is concerned, specifically the rights of translation, reprinting, reuse of illustrations, recitation, broadcasting, reproduction on microfilms or in any other physical way, and transmission or information storage and retrieval, electronic adaptation, computer software, or by similar or dissimilar methodology now known or hereafter developed.
The use of general descriptive names, registered names, trademarks, service marks, etc. in this publication does not imply, even in the absence of a specific statement, that such names are exempt from the relevant protective laws and regulations and therefore free for general use.
The publisher, the authors and the editors are safe to assume that the advice and information in this book are believed to be true and accurate at the date of publication. Neither the publisher nor the authors or the editors give a warranty, expressed or implied, with respect to the material contained herein or for any errors or omissions that may have been made. The publisher remains neutral with regard to jurisdictional claims in published maps and institutional affiliations.

This Palgrave Macmillan imprint is published by the registered company Springer Nature Switzerland AG.
The registered company address is: Gewerbestrasse 11, 6330 Cham, Switzerland

If disposing of this product, please recycle the paper.

To the memory of my parents

Preface

I.

The title of the book, *The Limits of Technology and the End of History*, underlines an intrinsic connection between the long-standing belief in the infinite potential of human mastery over nature and the Western ideal of man as a universal being and a subject of the universal world history. The ideologeme of infinite scientific and technological growth becomes a cornerstone of Marxist, pseudo-Marxist, and liberal philosophies. In particular, the founders of Marxism explicitly appeal to the "unbounded expansion of industry" as a driving force behind the transition from the 'prehistory' of socioeconomic strife to the harmony of social humanity; likewise, Fukuyama finds in technology a foundation of the continuing directional progress.[1]

In this context, any halt or a significant slowdown of technological progress would cause a return to the "old filthy business" (Marx) of socioeconomic and political struggle at the national and international levels. On the global scale, technological deceleration would amplify the reversal of the globalization trend by enhancing the existing and creating new geopolitical and economic division lines. Then, the 'end of history,' conceived by Marxists and like-minded (pseudo)liberals as the beginning of the *true history* in the 'realm of freedom,' would, instead, materialize as a

---

[1] Engels, F. (1976). "Principles of Communism" In Marx, K. & Engels, F. *Collected Works* (Vol. 6, p. 341). International Publishers. Fukuyama, F. (1992). *The End of History and the Last Man*. Free Press.

new 'state of nature,' with fierce global competition for resources, markets, and labor.

From this perspective, Western historical and technological optimism, in all its varieties, may represent, in Toynbee's words, only the "old parochial self-centered standpoint," inconsistent with an ambitious aspiration for political and cultural unification of the world.[2] This inconsistency is especially conspicuous in Marx's theory of 'scientific communism' that combines the close-ended dialectic of technological progress with the idea of open-ended (infinite) history.

The *epistemic* problematic of the Western philosophy of history is replicated in the problematic of its *value system* as a (generic for the modernity) tension between a productivist idea of progress and the conception of man as a *universal being*. At stake is a perennial existential choice of *to have* or *to be*, going back to the early Cartesian dualism of man as both a free, *thinking substance* and a "master and possessor of nature." In Marx's historical schema, the choice is between an objective of purging humanity from the "sense of having," the idea he advances already in the 1844 *Economic and Philosophical Manuscripts* and a conception of unbounded technological self-externalization and multiplication of human wants, the theme introduced in his early works and reinforced in the 'mature' *Grundrisse* and the *Capital*.

It hardly needs to be argued that modern technology is inseparable from the advancements in science. In this context, examining the epistemic foundations of the belief in infinite progress of (techno)science, we purport to show that *actual infinity*, either in the form of unbounded technological/scientific expansion or infinite complexity of nature, is redundant for the universality of man, his scientific pursuit and historical experience. What man, as an allegedly *universal being*, uncovers in the course of theoretical and practical appropriation of the world is a system of 'invariants of experience,' coextensive with human praxis. In Marx's anthropocentric ontology, this means that the universality of man emerges through labor practices, rendering nature "his extended body."[3] In the last analysis, the conundrum of universality and power calls, in Kant's terminology, for a systematic critique of *instrumental reason*, its practical applicability and value structure.

---

[2] Toynbee, A. J. (1948). *Civilization on Trial* (p. 83). Oxford University Press.

[3] Marx, K. (1973). *Grundrisse, Foundations of the Critique of Political Economy* (pp. 415–416) (M. Nicolaus. Trans.). Penguin.

II.

The book follows mostly a historical method of exposition. Within this frame of reference, we place the idea of infinite progress within the European cultural tradition and trace its origins from Cartesian rationalism to Hegelian dialectic of the Notion, its consequent 'inversion' in Marx's dialectic of technology-based labor, and 'beyond.' At the same time, this study should not be perceived as a history of ideas or a primer on Marx and Marxism. Marx's philosophy is explored to the extent that his dialectic of labor sheds light on Western technological optimism and the idea of universal history and provides an elaborate theoretical framework for analyzing the corresponding postulate of unbounded technological expansion with all its conceptual ambiguity. While Marx's philosophy is explored mainly in the context of his 'early' works, it is germane to note that the conceptual apparatus developed by 'early' Marx, especially in the *Economic and Philosophical Manuscripts*, provides an indispensable theoretical and humanistic underbelly for his later works, including the *Capital*. On this account, the widespread assumption of a radical break between the 'early' and 'mature' Marx is largely unjustifiable (and is not supported by this author) unless one concurs with Spengler's assessment of the Marxist socioeconomic model as a reverse form of capitalism.[4]

Chapter 1 (the Introduction) is a detailed overview of the book, with particular attention to the social and political ramifications of the idea of infinite growth. The sections, *From Theology to Omnipower of Reason* to *Hegel and the Idea of the State*, cover the material in Chap. 2. They address the evolution and ultimate crisis of European rationalism from Descartes to Hegel, providing historical and epistemological background for the consequent chapters. The following section, *From State Power to the Power of Historical Consciousness*, addresses the reevaluation of Hegel by his left-wing followers, who turned the closed-ended Hegelian dialectic into an instrument of radical social critique, the central topic of Chap. 3. The section *Marx's Dialectic of Technology and Infinite Progress* explores the central theme of Chap. 4: inconsistency between the circular structure of Marx's dialectic of labor and the postulate of infinite technological progress. Chapter 5 examines epistemological, historical, and social aspects of

---

[4] Spengler, O. (2013). *Prussianism and Socialism*. Isha Books.

the debates on the 'end of science' and 'limits to growth,' which are only briefly addressed throughout the Introduction.

Chapter 2 demonstrates how the incompleteness of theoretical attitude to the world emerges as a fundamental problem of modern philosophy and, through Cartesian rationalism, its Kantian transcendental critique and Hegelian dialectic, triggers a decisive turn to the understanding of philosophy as social praxis by Hegel's early nineteenth-century liberal critics, Marx included. In this context, we trace in what way the Cartesian interplay of knowledge and being finds a new form in the circular Hegelian dialectic of the Notion, providing a foundation for Marx's, allegedly materialist, dialectic of labor. As we argue in the two subsequent chapters, Marx's conception of labor, as both a *universal* and *material* mode of human activity, replicates the Cartesian circle of *formality* and *actuality* and, as a result, proves inconsistent with the ideal of infinite directional progress driven by the unbounded progress of science and technology.

Chapter 3 examines the transition from Hegel's philosophy of history to nineteenth-century political liberalism and its consequent critique by Karl Marx and Moses Hess. We discuss the disintegration of the Hegelian school, starting with a landmark theological critique of David Strauss and proceeding to the Fichteanized historical dialectic of Bruno Bauer, the liberal politics of Arnold Ruge, and the anthropocentric inversion of Hegel by Ludwig Feuerbach. The tension between the progressive spirit of the Enlightenment and Hegel's historical conservatism is conceptualized by the Young Hegelians as a contradiction between the *open-ended* nature of his revolutionary dialectical method vs. a *close-ended* structure of the speculative system and conception of history. The deconstruction of Hegelian philosophy, sharpened by Hess and the founders of Marxism, culminates in the theory of *socialism* and shapes Marx's conception of labor as a *universal* form of human activity and a principal instrument of socioeconomic emancipation.

The last two chapters present the core of this author's argument. Chapter 4 analyzes Marx's conception of labor in the framework of his *anthropocentric ontology*. We show how Marx substitutes an abstract Hegelian dialectic with an allegedly materialistic dialectic of self-reference and objectification, characterizing man as a being with "all his powers having the quality of self-reference."[5] The fundamental problem of such an

---

[5] Marx, K. (1971). *Economic and Philosophical Manuscripts of 1844* In Fromm, E. *Marx' Concept of Man: with a Translation from Marx's Economic and Philosophical Manuscripts*

'inverted' Hegelian methodology is the inconsistency between the circular pattern of the self-referential dialectic of labor with assumptions of infinite technological progress and open-ended history. To explicate this problem, we briefly discuss an alternative late nineteenth-century approach to technological evolution by the *organ projection theory*, where, by contrast with Marx's speculative dialectical schema, the methodological principles of German speculative philosophy are complemented by archeological, anthropological, and philological data.

In the second part of Chap. 4, we proceed 'beyond Marx' and consider the early twentieth-century developments in orthodox and revisionist Marxism. The chapter concludes with a discussion of an unorthodox interpretation of Marx by the underrated Russian Marxist activist, scientist and philosopher A. Bogdanov. In his highly original synthesis of *empiriocriticism* and Marxism, the materialist dialectic is construed in the framework of the *general theory of organization*, viewed by some as an early precursor of *cybernetics* and *general systems theory* (*Tektology*). The innovative spirit of Bogdanov's philosophy and, in particular, his attempted *tektological* proof of "inexhaustible human creativity" resonate with the major developments in twentieth-century science. However, his eclectic synthesis of Ernst Mach and Karl Marx reveals a gaping chasm between the allegedly scientific shell and the Marxist ideological underbelly of his philosophy of history.

Chapter 5 begins with a discussion of Bogdanov's *general theory of organization* and his innovative proof of inexhaustible human creativity and proceeds to examine the contemporary 'end of science' and 'limits to growth' debates. Examining a wide range of issues in epistemology, the philosophy of science and the foundations of mathematics, we try to show that the potential limits to scientific and technological progress are typically explored in the heteronomous frameworks of practical constraints external to its intrinsic logic. At the same time, the (philosophically naïve) arguments from the allegedly theoretical perspective are usually construed within a narrowly defined branch of fundamental science and are either incomplete or inconsistent. What is required is a meta-scientific, epistemic analysis of scientific/technological progress, its qualitative scope, and temporal extent.

The Coda elaborates on the key issues discussed throughout the book, further explicating a connection between epistemological, social, and

(pp. 178–179) (T. B. Bottomore. Trans.). Frederick Ungar Pub. This work is often referred to as *Paris Manuscripts*.

ethical aspects of technological and scientific progress, including the impact of the belief in infinite growth on technological practices. As we argue throughout the book, technology-based labor does not represent a universal form of human activity and the genuine synthesis of theory and practice, culture and nature cannot be accomplished through the unceasing technological externalization of human "essential powers" (Marx). In this context, we contrast the Western productivist idea of history with an alternative historical *telos* of "the maximum of well-being with the minimum of consumption."[6] Answering Kant, what we, at least, *cannot* hope for is the unbounded expansion of human mastery over the world.

III.

The book is conceived as an interdisciplinary study that may appeal to all (including undergraduate and graduate students) interested in the history of modern philosophy, Marx and Marxism, the philosophy of history, political liberalism, and the philosophy of science/technology. The footnotes and the Bibliography are intended to guide the reader in further research. The Bibliography includes referenced sources, alternative editions, or translations of the cited works and potentially helpful complementary resources.

While the author tried to present a consistent line of argument throughout the book, from introductory remarks in this Preface to the final historical and epistemic assessments in the Coda, some readers, based on their interests and background, may be able and willing to proceed from the introductory overview in Chap. 1 to one of the consequent chapters: Chapter 2 for the historical and epistemological analysis of seventeenth-eighteenth century European rationalism, Chap. 3 for post-Hegelian philosophy and early Marx, Chap. 4, for Marx's philosophy of technology and the early twentieth-century Marxism, or Chap. 5, for the analysis of contemporary limits to science and technology debates; again, the last two chapters (including the Coda) are essential for the understanding of the author's argument.

I want to thank the Palgrave Macmillan publisher, Mr. Brendan George, the editors, Ms. Robin James, Ms. Amy Invernizzi and Mrs. Sujatha Mani, and the rest of the editorial team for their support and assistance with

---

[6] Schumacher, E. F. (1999). *Small is Beautiful: Economics as if People Mattered* (p. 40). Hartley & Marks.

publishing this book. I am also grateful to the anonymous reviewer for thoughtful reading and helpful suggestions. Finally, I want to complement a chapter four discussion of 'orthodox' Marxism and ideological and political developments in modern Russia by expressing my support for men and women of conscience, persecuted for the "courage to use their own reason"(Kant), including the recently imprisoned director Y. Berkovich, playwright S. Petriychuk, and political scientist B. Kagarlitsky.

Plymouth Meeting, PA, USA							Yefim Kats

Contents

1 **Introduction**   1
   From Theology to Omnipower of Reason   3
   The Paradox of Knowledge and the Idea of History   4
   Human Freedom Reclaimed   6
   Hegel and the Idea of the State   7
   From State Power to the Power of Historical Consciousness   9
   Dialectic of Technology and Infinite Progress   11
   The End of History and the Prospects of Unbounded Growth   13

2 **The Dialectic of Knowledge and the Idea of History**   19
   In the Circle of Knowledge and Reality   22
   The Limits of Reason and Dualism of History   29
   History as a Triumph of Freedom   36
   The Dialectic of History and the Myth of the State   41
   From Dialectic of Reason to Historical Praxis   45

3 **From Social Anthropology to the Historical Dialectic of Labor**   49
   From Divine Providence to Humanity   52
   The Dialectic of Reason as Social Critique   59
   From Social Critique to Social Anthropology   64
   Social Anthropology as Socialism   68
   Emancipation of Human Senses and Ontology of Labor   74

4   The Dialectic of Labor and the Limits of Technological
    Growth   79
    The Marxian Circle: Labor, Technology, and Nature   81
    Infinite Growth and Dialectic of Self-Externalization   89
    Marxist Orthodoxy and Beyond   98
    Towards Dialectical Materialism   103
    Philosophy of Praxis Vs. Dialectical Materialism   106

5   Towards the Critique of Technological Reason   113
    From Philosophy of Praxis to General Theory of Organization 116
    Inexhaustible Creativity and Infinite Growth   118
    Technology and the Limits to Growth   121
    The End of Science and Invariants of Human Experience   125
    Coda: Knowledge, Freedom, and the Riddle of History   136

Bibliography   147

Name Index   161

Subject Index   165

CHAPTER 1

Introduction

> *"Modern Times."* A story of industry,
> of individual enterprise—humanity cruising in the
> pursuit of happiness.
> —Chaplin

Throughout the twentieth century, Marxists viewed the upcoming communist utopia as the end of prehistory and the beginning of the historical trajectory of freedom. On the other side of the 'iron curtain,' the broadly conceived idea of liberal democracy was advanced as an alternative 'end' of history and ideology, ushering in a new era of political unification of the world. At the end of the century, the tectonic geopolitical shifts seemingly came as a triumph of liberal worldview, aiming to encompass a diversity of world cultures in a universal political framework, just "one station behind communism."[1] The prophetic universalism of the Western historical

---

[1] Fukuyama, F. (1992). *The End of History and the Last Man* (Chap. 5). Free Press and Fukuyama, F. (2007, July 2). "The End of History Revisited." Seminars About Long-Term Thinking: https://www.youtube.com/watch?v=w240nD5whsE. The similarity with Marxism is reinforced by Fukuyama's assessment of the Hegelian idea of the State as an early blueprint of liberal democracy. For Marx's 'solution of the riddle of history,' see Marx, K. "Economic and Philosophical Manuscripts" In Marx, K. (1975). *Early Writings* (trans. by R. Livingstone and G. Benton) (p. 348). Vintage Books.

© The Author(s), under exclusive license to Springer Nature Switzerland AG 2024
Y. Kats, *The Limits of Technology and the End of History*,
https://doi.org/10.1007/978-3-031-69545-2_1

outlook is clearly articulated by Spengler, who emphatically declares that Western "man differs from all others in his insatiable will to reach the infinite.... He uses his historical thinking to take hold of the past and integrate it into his own existence under the name of 'world history.'"[2]

To a large extent, the Eurocentric idea of universal world history has been driven by the historical and technological optimism of the Enlightenment. In Spengler's words, in the Western world, "technics is eternal and immortal like God the Father, it delivers mankind like God the Son, and it illumines us like God the Holy Ghost. And its worshiper is the progress-philistine of the modem age which runs from Lamettrie to Lenin."[3] More than 50 years later, Fukuyama as well refers to the continuous scientific and technological expansion as an instrument of "directional" progress and reiterates the Cartesian idea that "man's superior dignity entitles him to the conquest of nature… made possible through modern natural science."[4] The faith in the "unbounded expansion" of industry is also a vital element of the Marxist conception of history. The founders of Marxism claimed that the whole of human history had been the history of industry and continuous technological progress was a necessary precondition for the transition to and the sustainable development of a new historical epoch of freedom.

As the omnipower of reason replaces divine power, the idea of man as a universal being finds support in the unshakable faith in the infinite progress of science and technology, molding an environment of ever-growing material wants. Reshaping not only the fabric of social life but pertaining to the very nature of man, such an environment potentially transforms man from a universal *human being* to a universal *consumer being*, depriving society of its humanistic core, "its total 'Existenz.'"[5]

In order to understand the epistemological, moral, and historical implications of the cultural crisis associated with a secular religion of infinite growth, in the following sections, we trace its roots to the dawn of modern philosophy, from the Cartesian project of mastery over nature to the prophetic Marxist and liberal forms of historicism based on the persistent

---

[2] Spengler, O. (2013). *Prussianism and* Socialism (p. 19). Isha Books (reprint of the 1920 edition).

[3] Spengler, O. (1976). *Man and Technics: A Contribution to a Philosophy of Life* (p. 43) Greenwood Press (reprint of 1932 ed. by Alfred A. Knopf, Inc.).

[4] Fukuyama, F. (1992). *The End of History and the Last Man*, p. 297.

[5] Husserl, E. (1970). *The Crisis of European Sciences and Transcendental Phenomenology* (p. 12). Northwestern University.

faith in technology-driven directional progress. We analyze the epistemological and social issues associated with the belief in unbounded technological progress and purport to show that infinite growth is inconsistent with Marx's circular dialectic of labor and represents hardly more than a doctrinal assumption embedded in the high Renaissance and Enlightenment. This position is further supported by the analysis of the 'end of science' and 'limits to growth' debates addressed in the last chapter.

FROM THEOLOGY TO OMNIPOWER OF REASON

European historical universalism finds a brilliant exponent in Renè Descartes (1596–1650), often considered the 'father' of modern philosophy.[6] The progressive pathos of Cartesian rationalism signifies a decisive breakup with the theological way of thinking that gave a powerful impetus to the historical optimism of the Enlightenment. At the same time, the fundamental mythologeme of infinitely expanding human mastery over nature gets firmly embedded in the European historical outlook.

Descartes advances the idea of man as a *universal being*, distinguished by the unique ability of rational reasoning. However, while his novel methodology is focused on the creative power of reason, this power extends only to the epistemic certainty, with God, in a tribute to the scholastic philosophy, viewed as a creator and sustainer of the universe. The divide between subjective certainty and objective reality encompasses even man, conceived by Descartes as a rational thinking substance trapped in the bodily machine.

In the strict limits of Cartesian rationalism, nature is conceptualized as a *mathematical manifold*, and mathematics is considered a model of rational thinking. The mathematical invariability (universality) of the laws of nature allows man and nature to enter into a relationship of *universal unity*. In this context, Descartes envisions the "project of universal science which can elevate our nature to the highest degree of perfection."[7] As he proclaims in *The Discourse on Method*, "Instead of that speculative philosophy which is taught in Schools... we can find a practical philosophy by means of which... we can render ourselves the masters and the possessors

---

[6] Cassirer, E. (1965). *The Philosophy of the Enlightenment* (p. 158). Beacon Press.
[7] A tentative title of his masterpiece, *Discourse on Method* (1637).

of nature."[8] The unlimited power of science is conceived as an engine of progress and a principal tool for the amelioration of the human condition, while an assumption that the *universality of natural laws implies that science can progress indefinitely* and secure infinite progression toward human mastery over nature becomes a guiding idea for the consequent development of European thought, from Kant's transcendental critique to Marx's dialectic of labor.[9]

Despite its progressive historical spirit, the rigorous Cartesian rationalism was challenged by empiricist critique (most notable by Hume) and advances in experimental science. On this background, Gottfried Wilhelm Leibniz (1646–1716) offers a more balanced account of the relationship between spirit and nature, underlining that the world has 'no hiatus.' The problematic of Cartesian rationalism, Leibniz's epistemological reflections, and his idea of harmony between the physical world and the world of grace open the way to the Kantian Copernican revolution in philosophy, with its conception of history as an infinite progression toward an ultimate unity between the natural 'realm of necessity' and the moral 'kingdom of ends.'

THE PARADOX OF KNOWLEDGE AND THE IDEA OF HISTORY

Kant's (1724–1804) aspiration is to reconcile the Cartesian divide between the assumed universality of man as a rational agent and the contingency of human material existence. Characterizing the spirit of the Enlightenment, he emphatically proclaims: "Have the courage to use your own understanding; this is the motto of the Enlightenment."[10] This moral dictum finds a vivid expression in the *transcendental* principle of Kant's critical philosophy: all we may hope to know is what we can *produce* through the constructive activity of reason.

While the idea of man as a free rational agent remains a guiding principle of Kant's architecture of reason, his picture of the world retains the

---

[8] Descartes (1978). *Discourse on Method* (Part VI) In Descartes. *The Philosophical Works of Descartes* (E. S. Haldane and T. R. G. Ross, Trans.) (Vol. 1). (p. 119). Cambridge University Press. See also Schouls A. P. (1991). "Descartes and the Idea of Progress." In *Rene Descartes: Critical Assessments* (J. D. Moyal. Ed.) (Vol. 1) (pp. 50–60). Routledge.

[9] Bury, J. B. (1920). *The Idea of Progress: An Inquiry into Its Origin and Growth*. (p. 66). Macmillan and Co.

[10] Kant. "Beantwortung der Frage: Was ist Aufklanrung?" In Kant. *Werke* (E. Cassirer. Ed.). (Vol. 4) (p. 169).

elements of Cartesian dualism, albeit in a more sophisticated form. Both nature as it is *in itself* and man *in himself* as a (transcendental) subject of freedom and moral judgment remain beyond the scope of *theoretical reason*. Instead, human freedom manifests itself through the moral imperatives of *practical reason*, independent of either natural necessity or God's will. In the realm of nature, "human actions are determined, like every other natural event, by the universal laws [of nature]," while the autonomy of *free will* empowers the universality of moral law. Thus, in Kant's universe, reality is split between natural necessity and moral freedom, and human reason is divided between its theoretical and practical modes.[11]

Hence, the paradox of the Kantian conception of history where historical progress, driven by the very dualism of necessity and freedom, can be accomplished only through the infinite convergence of "human natural capacities" to the moral 'kingdom of ends.' The unity between the natural world and the 'kingdom of ends' is to be implemented through the "steady and progressive though slow evolution of human species," which is possible only within the tenets of "civil society," where its members enjoy the highest freedom.[12] In civil society, the gradually abating antagonism between individual inclinations and the requirements of moral law assumingly leads to the development of "human natural capacities" according to "the secret plan of nature."[13]

In line with the dual character of historical progress, Kant introduces a normative distinction between the humane *culture of discipline* and the instrumental *culture of skill*. The leading role in history belongs to man, with his unique ability to separate the categorical requirements of moral law from the survival skills used in his natural environment. Kant argues that the uncompromising commitment to follow moral requirements constitutes a genuinely humane culture; by contrast, the culture of skill should play only a supporting role in the progression toward the final synthesis of human freedom and natural necessity. Due to the intrinsic dualism of spirit and nature, moral law and natural inclinations, such synthesis remains only a normative requirement for the *infinite* "history of human species."[14]

---

[11] Kant "Uber die von der Konigl…" 1, 2, 2.
[12] Kant, I. (1784). *Idea for a Universal History from a Cosmopolitan Point of View* In Kant. (1963). "On History," The Bobbs-Merrill Co.
[13] Kant. *Critique of Judgment*, Par. 83.
[14] Kant. I. *Idea…* Par. 8.

On this background, the pressing point for the consequent development of European philosophy was to bridge the gap between freedom and necessity, the moral telos of the historical process and its material foundation, and understand man in the integrity of his individual and historical existence.

Human Freedom Reclaimed

Kant found a gifted and enthusiastic supporter in his younger colleague, Johann Gottlieb Fichte (1762–1814), whose philosophy emerges as an epitome of the Enlightenment hymn to human freedom and historical progress. Reinforcing, in the letter to K. L. Reinhold, the moral core of his philosophy, Fichte proclaims that his "system from the beginning to the end is only an analysis of the notion of freedom." He sharpens the progressive spirit of Kantian transcendental methodology, affirming the ultimate primacy of human freedom and autonomy over the necessity of natural inclinations.[15]

The leading principle of Fichtean philosophy is the activity of the absolute I. In its free historical unfolding, the I represents a synthesis of activity and receptivity, freedom and necessity.[16] In this framework, nature, construed as a byproduct of the spontaneous activity of the I, is effectively reduced to an external horizon of the perpetually expanding spiritual Big Bang, providing a material foundation of human progress. Paraphrasing Mill, nature (matter) appears as a permanent possibility of subjective activity, a passive (material) vehicle for infinite self-realization of the absolute I.

Due to its subordinate status, the man enters into purely instrumental relations with nature, utilizing it as an *inexhaustible* medium for the satisfaction of his needs. A true son of the Enlightenment, Fichte emphasizes the vital role of scientific "discoveries and inventions" in historical progress and the continuous amelioration of the human condition: "The satisfaction of human needs must become easier and easier… there must be made an infinite number of new discoveries and inventions [underlining

---

[15] An outline of these ideas is in Гайденко, П. П. (1979). *Философия Фихте и Современность*. Издательство Мысль. (Gaidenko, P. P. *Fichte's Philosophy and Modernity*). A brief but thoughtful review of German philosophy from Kant through Hegel is in Copleston, S. J. (1963). *A History of Philosophy* (Vol. 7). Image Books.

[16] With an implicit reference to Leibniz, Schelling, Fichte's follower in his early period, refers to I as a completely closed in itself *monad*. Schelling, F. W. J. (1978). *System of Transcendental Idealism*. trans. by Peter Heath, Charlottesville: University Press of Virginia.

throughout the book by is YK] to increase the means of subsistence."¹⁷ The open-ended conception of history and the paradigmatic idea that scientific and technological advancements are the principal contributors to human progress became key elements of European ideology, firmly incorporated in Marx's prophecy of 'scientific communism.'

Though Fichte insists that the body of the historical process and its moral core constitute a unified whole instead of belonging to ontologically different realms (as Kant thought), an abstract ontology of nature remains the Achilles heel of his historical dialectic. As a result, in his philosophy of history, humanity, as a vehicle of the absolute I, is destined to continuously transcend the duality of its earthy existence in the infinite strive toward the elusive kingdom of ends. The abstract approach to nature and the gaping chasm between the moral goals and material basis of history became a target of critique for his celebrated colleagues, Schelling and Hegel, and for Kant himself.

HEGEL AND THE IDEA OF THE STATE

Trying to resolve the recurrent dualism of European rationalism, G. W. F. Hegel (1770–1831) emphasizes that reality should be conceived as a *process* of dialectical *becoming* of the Absolute, which constitutes "the circle that presupposes its end as its goal and has its end for its beginning, and which is actual only through this accomplishment and its end."[18] In the historical self-realization of the Absolute, the initial undifferentiated *in itself* identity of the matter and spirit is transformed into the logically structured *for itself* totality. This process is driven by the *formal* (in the scholastic sense) objectivity of dialectical laws.[19] In this way, Hegel, following Kant and Fichte, further generalizes the principle of Cartesian *cogito* by replacing intellectual intuition with the *formal* (and, hence, 'real' in the scholastic tradition) laws of thought.

---

[17] Fichte, J. G. (1794). "Einige Vorlesungen über die Bestimmung des Gelehrten." In Fichte, I. G. (1845–1846). *Sammtlihe Werke*. (Fichte I. H., (junior). Ed.). (Vol. 6, p. 342).

[18] Hegel, G. W. F. (2018). *Phenomenology of Spirit*. (p. 12). Cambridge University Press.

[19] In Chaps. 3 and 4, we explore an alternative version of dialectic developed by one of the leading twentieth-century Marxists, A. Bogdanov.

In its historical unfolding, the Absolute ultimately emerges in the *concrete* form of "universal world spirit," represented by the State.[20] Hegel emphasizes that the State is superior to *civil society*, which is, independent of its political makeup, a quasi-free association of isolated individuals incapable of expressing what Rousseau calls the *general will*. On this account, a democratic consensus of singular individuals in civil society does not reflect a true spirit of human freedom and does not represent humanity in its essential unity. Hegel argues that the relationship between an individual and the State, as an embodiment of freedom, should be entirely different: the truly wholesome and free individual existence is realized only through State membership and supersedes any contingent contractual agreement, including democratic forms of governance; accordingly, the scope and extent of individual freedom are to be determined exclusively by the State.[21] Considering that for Hegel, constitutional monarchy is the highest form of the State, it is not a surprise that in the Kingdom of Prussia, his homeland, Hegelian philosophy attained the status of a virtually official state ideology.

A wide range of Marxist and pseudo-liberal historical models (including Fukuyama's) are conspicuously linked to the Hegelian state-focused political ideology and its descendants. In this respect, Ludwig von Mises, a champion of 'classical' liberal tradition, observes that modern liberal policies "originated in Europe and... their most brilliant nineteenth-century exponent was Bismarck... Bismarck's Sozialpolitik was inaugurated in 1881, more than fifty years before its replica, F. D. Roosevelt's New Deal."[22] He further argues that the currently prevailing forms of liberalism exemplify a potentially oppressive political *statism* (or *etatism*). Along with Sozialpolitik, such forms of liberalism could be indeed traced to the Hegelian conception of the State and is more akin to what Spengler calls *Prussianism* than to the practice of liberal democracy in civil society.

Unlike Kant and Fichte, Hegel strongly opposes the idea of what he labels a 'bad infinity' of historical progress. This position, and his overall

---

[20] Hegel, G. W. F. (1927). *Philosophische Propadeutik* (Par. 202) In Hegel, G. W. F. *Samtliche Werke* (Bd. 3). Hrsg. Von H. Glockner. See also Hegel, G. W. F. (1942). *Philosophy of Right* (Introduction). Clarendon Press.

[21] Hegel criticizes Rousseau and Fichte for the alleged understanding of the general will as a consensus of isolated individuals. See Hegel, G. W. F. (1942). *Philosophy of Right*, III, 3 (State), par. 257–258.

[22] Mises, L. (1985). *Liberalism in Classical Tradition* (3rd ed) (p. 18). The Foundation for Economic Education & Cobden Press (online 2002 ed. by Mises.org)

political conservatism, provided a foundation for the claims that, for Hegel, the end of history "is not only attainable but has now been attained" or that his philosophy of history is "antagonistic to [future] progress" and indifferent to the "amelioration of the material condition of human life."[23] The controversy over Hegel's political conservatism and the close-ended conception of history (vs. his assumingly open-ended dialectic) triggered a sharp split between his followers while providing fertile soil for a wide range of historical, social, and political ideologies.

From State Power to the Power of Historical Consciousness

After Hegel's untimely death in 1831, an initial controversy over the relationship between his system and Christianity outgrows into a fierce debate over the philosophy of history. At stake was the center of historical gravity—humanity vs. the Absolute (God), the democratic consensus of civil society vs. the Orwellian power of the State.

In this debate, an important role belongs to Hegel's left-wing critic, Ludwig Feuerbach (1804–1872). The essence of his anthropological interpretation of Hegel is succinctly expressed by the motto—"the secret of theology is anthropology."[24] From this standpoint, the idea of God (hidden under a speculative guise of the Absolute) is a reflection of an objectified, by imagination, essential human nature, and the assumed universality of God is nothing but a concealed universality of man, conceived in the totality of his *species existence*. In the polemic with Max Stirner, Feuerbach explicitly states that in his anthropology, "the predicates of God... become the predicates of nature and mankind," and speculative philosophy is "brought from heaven to earth." This secularization of the Hegelian system paved the way for the subject-oriented idea of history and the conception of philosophy as social critique.

While Feuerbach primarily attacked the theological core of Hegel's system, Bruno Bauer (1809–1882), a theologian turned a fierce opponent of theology, offered a 'Fichteanized' version of the Hegelian dialectic by targeting an alleged contradiction between the open-ended method and the close-ended structure of Hegel's system and his philosophy of history.

---

[23] Bury, J. B. (1920). *The Idea of Progress,* pp. 255–256.
[24] Feuerbach, L. (1957). *The Essence of Christianity* (George Eliot. Trans.). Harper & Row Pub.

Bauer challenges the two key assumptions of Hegelian philosophy. First, according to Bauer, the dialectic of reason entails that no concrete historical phenomenon could be final; in other words, no 'end of history' is possible, and no particular social formation (such as the state) can represent an embodiment of ultimate freedom. This critique applies, in particular, to Hegel's claim that constitutional monarchy is the highest form of government in Prussia and, more broadly, for the Germanic world.

Second, Bauer rejects Hegel's contention that human consciousness does not play an active role in the grand historical drama and that people only retrospectively reflect on the historical unfolding of the Absolute. He argues that the power of philosophical critique is a driving force in advancing moral and social freedom and plays a decisive role in the progressive historical transformation toward the kingdom of moral ends. From this perspective, it is a radical reinterpretation of Hegel by Bauer and his circle that could be viewed as an early form of political liberalism, while appeals to Hegel as a precursor of liberal ideology may reflect only a distorted idea of what the genuine liberal philosophy is about.

Despite the progressive spirit of Bauer's critique, he presents a distorted, 'Fichteanized' form of the Hegelian dialectic, and his historical optimism is more an article of faith than an outcome of the elaborate critical analysis of Hegelian philosophy. Moreover, while Bauer claims that the power of philosophical critique represents a principal instrument of social and political change, he fails to tie it to the concrete socioeconomic condition of human life. As a result, his philosophy recapitulates the Fichtean disconnect between the ideal mode of human activity and the material mode of human existence.

The particular importance of the left Hegelians lies in their influence on the development of liberal ideology and the progressive European social and political movement of the nineteenth century. It suffices to say that Karl Marx and Friedrich Engels, in their early career, were associated with the group.[25] At the center of Marx's prophetic conception of history and the view of man as a *universal being* is his idea of labor as a *universal* mode of human activity. However, Marx's strife for human universality results in a major methodological problem: an inconsistency between the idea of open-ended historical progress (he shares with Bauer and Fichte) and his essentially Hegelian, close-ended dialectic of technology-based labor.

---

[25] Marx, K. "Theses on Feuerbach." A detailed discussion of the Young Hegelians and their impact on Marx's thought is in Chap. 2.

Dialectic of Technology and Infinite Progress

While Bauer and his friends de facto rejected not only Hegel's system but also the very core of his dialectic and, instead, turned to Fichte as their inspiration, Marx aspired to combine the 'inverted' Hegelian essentialism with the open-ended spirit of Fichte's conception of history. In his view, such synthesis would elevate philosophy to the level of *humanism*, surpassing both the contemplative materialism of Feuerbach and the abstract idealism of Hegel and Bauer.[26] At the center of Marx's program is a contention that the genuine mode of universal (aka objective aka "practical-critical") activity commensurate with man as a universal being is *productive labor*. His ultimate goal is to understand man as a *total* being, in the unity of *material* and *ideal* forms of life.

Technology plays a fundamental role in Marx's dialectic of labor. Due to the mediating function of technology in human interaction with nature, man is capable of *objectifying* himself *for himself* in his natural environment, thus becoming a *universal natural being*. In the assumed inversion of Hegelian methodology, Marx's dialectic of labor de facto employs the *in-itself* to *for-itself* pattern of the dialectical unfolding of the Absolute under the guise of a dialectic of human *self-reference* and *objectification*: "**Man is self-referring… Every one of his faculties has this quality of self-reference**… [and]… the **real**, active orientation of man to himself as species-being (i.e. as a human being) is possible so far as he really brings forth all his species-powers… and treats these powers as objects."[27] It is the self-referential character of the "active" orientation to the world that allows man to reveal *for himself* (in the *objective* form) his own physicality and, "consequently" (as he emphasizes in the *German Ideology*) that of nature.[28] In the spirit of German transcendental philosophy, he maintains that the *objective* mode of nature appears to man only through the *objective* activity of technology-based collective labor. At the same time, man

---

[26] Popper, while praising Marx's "activism" aimed at social change, indicates that his "essentialism" is due to the heavy influence of Hegel and Plato. See Popper, K. (1966). *The Open Society and Its Enemies* (5th ed.) (Vol. 1, 2.). Princeton University Press. Underlying the Hegelian roots of Marx's methodology, Lenin remarks that one cannot understand *Capital* without a thorough knowledge of Hegel's *Science of Logic*.

[27] Marx, K. (1975). *Economic and Philosophical Manuscripts* In Fromm, E. *Marx' Concept of Man: With a Translation from Marx's Economic and Philosophical Manuscripts* (pp. 178–179) (T. B. Bottomore. Trans.). Frederick Ungar Pub.

[28] Marx, K. (1977). *The German Ideology* In Karl Marx *Selected Writings* (McLellan. Ed.) (p. 160). Oxford University Press.

also emerges *for himself* as an *objective* (universal) being *coextensive* with his "expanded body" of nature. Thus, while in the Hegelian panlogism, the ontology of the Absolute recapitulates "genetic exposition of the notion," in Marx's philosophy of labor, *ontology recapitulates technology*.

In his treatment of technology, Marx closely follows Hegelian dialectic. While for Hegel, tools represent only a 'moment' in the *close-ended*, self-referential dialectic of the Absolute, for Marx, the whole of human history is turned into the process of technological self-externalization.[29] Summarizing Marx's dialectic of technology, Tucker says, "Taking his clue from Hegel, Marx says that the history of production is an Entäusserungsgeschichte, a history of man's own self-externalization."[30] In this conceptual framework, tools, instruments, and the technological infrastructure, in general, represent externalized products of human 'essential powers,' and historical progress unfolds "through the co-operative endeavors of mankind," as the history of human technological self-externalization.[31]

In a sweeping conceptual generalization, Marx pronounced the whole of preceding human history as the "history of industry," claiming that for a successful transition to communism, the forces of production must continuously expand beyond the 'realm of necessity.' A failure to secure "unceasing and global" material growth may cause society to relapse into the "old filthy business" of socioeconomic struggle and suffering. Reinforcing this technological determinism, Engels claims that a fundamental precondition of communism is "the infinite possibility inherent in machinery, chemical and other resources of their unlimited growth."[32]

The fundamental problem of Marx's ontology of labor is that the circular character of his dialectic of technological self-externalization is *inconsistent* with the idea of unbounded technological growth and the open-ended conception of history. Furthermore, the 'bad infinity' of technological self-externalization is not only *inconsistent* with Marx's dialectic of labor but also logically *redundant* to his conception of man as a *universal being*.

[29] Hegel, G. W. F. *The Science of Logic*. 3, 2, 3.
[30] For the discussion of technology as a product of human self-externalization, see Chap. 4. Tucker, R. C. (1961). *Philosophy and Myth in Karl Marx* (pp. 130–131). Cambridge University Press.
[31] Marx, K. *Economic and Philosophical Manuscripts*, pp. 178–179.
[32] Engels, F. "Draft of a Communist Confession of Faith" and "Principles of Communism" In Marx, K. & Engels, F. (1976). *Collected Works* (Vol. 6) (pp. 96, 341). International Publishers.

Instead, in the spirit of his anthropocentric ontology, the universality of man as a natural being is constituted by the *isomorphic relationship* between human physicality and nature.[33]

THE END OF HISTORY AND THE PROSPECTS OF UNBOUNDED GROWTH

The faith in the unlimited potential of technological appropriation of the world is shared across a broad spectrum of Western ideologies, from Marxist orthodoxy to its reformist adversaries and the proponents of liberal democracy.

An important role in the political transition from Marxism to liberalism belongs to Edward Bernstein's (1850–1932) theory of 'evolutionary socialism.'[34] Bernstein's famous motto, "the movement is everything, the final goal is nothing," implies that the working class should engage in peaceful politics focused on the gradual expansion of parliamentary representation and social welfare. The program of day-to-day political struggle within the limits of civil society effectively transforms Marxism into an ideology of liberal democracy, with man forever bound to the industrial 'prehistory' and no real ground left for the historic leap into the free realm of *social humanity*. At the same time, the idea of man as a *universal being*, with the proletariat as its social alter ego, loses its socioeconomic underbelly, and man is de facto turned into a *universal consumer being*. Suspended between the uncertainty of his socioeconomic reality and the promised land of an elusive bright future, he is entangled in the net of material production and expanding, to the 'unlimited extent,' 'wants.' It is an emphasis on the unbounded industrialization, accumulation of wealth, and expansion of human wants that allows Oswald Spengler (1880–1936) to assess Marxist models of (consumer) socialism as *reverse capitalism* based on the "expropriation of expropriators and robbery of the robbers."[35]

While a market-based liberal democracy distinguishes itself (from socialism) by allegedly not setting the "aim at creating anything but the outward preconditions for the development of the inner life," de facto, the

---

[33] The relationship between the postulate of infinite growth and the idea of man as a universal being is discussed throughout Chaps. 4 and 5.

[34] Bernstein, E. (2015). *Evolutionary Socialism*. Forgotten Books (reprint of 1899 German edition).

[35] Spengler, O. (2013). *Prussianism and Socialism* (p. 76). Isha Books.

unlimited material expansion and proliferation of 'wants' undermine the alleged independence between the "outward" conditions of human life and the "highest and deepest in man" that "cannot be touched by any outward regulation."[36] The expanding wants gradually suppress the genuine autonomy of 'inner life,' substituting it with the utilitarian ideal of (pseudo)happiness in the brave new world of 'affluent society.'[37] The persistent strive for material prosperity inevitably translates into the set of overriding social, economic, and moral objectives distorting the fragile autonomy of 'inner' life. Under these conditions, any halt or even a long-term deceleration of technological advancements would leave 'consumer democracies' in a 'natural state' of war over resources, labor, and markets, completely derailing the Enlightenment ideal of scientific/technological progress as a sure path toward the kingdom of moral and social ends.

The problem is only aggravated by the fact that, in the consumer society, liberal policies are guided by "those who succeed in making themselves accepted as a majority," including the corporate, party, and state establishment.[38] In such an environment, divisions along the party lines are often secondary to the actual power hierarchy, and democratic governance is more a matter of the balance of power and makeup of the elites than the extent of the political influence of the masses. Indeed, one would be hard-pressed to argue that in modern Western society, the socioeconomic equilibrium is being maintained as a liberal consensus of small business, working class, and labor leaders: consider the rise of corporate giants, class polarization, the demise of the labor unions, and global outsourcing.

The global distribution of power follows a similar pattern, with the self-designated representatives of 'accepted majorities' shaping an international agenda. At the same time, the mounting problems of globalization are routinely attributed to unintentional mistakes of the elites, endowed with an exclusive authority to design and implement any corrective policies. In his 1950th analysis of global power struggle, C. W. Mills remarks that the "privileged nations condemn weaker ones in the name of internationalism, defending with moral notions what has been won by force

---

[36] Mises, L. (1985), pp. 27 — 28.
[37] Huxley, A. (2005). *The Brave New World* and *Brave New World Revisited*. Harper Perennial Modern Classics. This problematic is also briefly discussed in Putnam, H. (1987). *The Many Faces of Realism* (pp. 57–61). Open Court.
[38] Mill, J. S. *On Liberty*, ch.1.

against those" who "can hope to change status quo only by force."[39] The current geopolitical environment bears a sad witness to the continuing relevance of such an assessment.

Meanwhile, the popular mythology of the upcoming liberal 'end of history' is routinely enforced by propaganda, "penetrating… deeply into the details of life, and enslaving the soul itself" by relentlessly pushing the utilitarian ideal of a happy life within a consumer society.[40] In the age of manufactured consent, "(1) the media tell the man in the mass who he is—they give him identity; (2) they tell him what he wants to be—they give him aspirations; (3) they tell him how to get that way—they give him technique; (4) they tell him how to feel that he is that way even when he is not—they give him escape."[41] In effect, the propaganda machine effectively molds the members of civil society into 'universal consumers' increasingly absorbed in the production/consumption loop and accumulation of wealth, whether on an equal or (pseudo)meritocratic basis. The attempts at direct political appeal to the people, especially accompanied by the critique of ruling elites, are considered a characteristic feature of populism.[42]

For the social mythology of unbounded consumerism, technology becomes a panacea for socioeconomic emancipation and a driver of "continuous and indefinite growth."[43] Even the pessimistic Western historical scenarios traditionally retain the long-standing hope of global growth based on infinite technological progress without ever questioning its sustainability. Already in the early twentieth century, at the time of perhaps the first globalization crisis, the prophesied by Spengler, the 'decline of the West' was construed in terms of cultural constraints rather than the intrinsic limitations of technological progress itself. Spengler's model of history assumes a recurrent cyclical shift of economic growth and technological innovations to new centers of power. More recently, Huntington (2011) and Wallerstein (2004) focus predominantly on the cultural clash of 'civilizations' or complex socioeconomic 'world systems' instead of the

---

[39] Mills, C. W. (1956). *The Power Elite* (p. 246). Oxford University Press.

[40] Mill, J. S. *On Liberty*, ch.1.

[41] Mills, C. W. (1956). *The Power Elite* (p. 313). Oxford University Press.

[42] Fukuyama also seems to follow this route. See Fukuyama, F. "The End of International Liberal Order?" Taipei, 04/17/2017: https://www.youtube.com/watch?v=scAzukYHJjY&list=PLsosaXrD-IswaVIYiygI6ZqYWnlKwin5P&index=45

[43] Gendron, B. (1977). *Technology and the Human Condition* (p. 24). St. Martin's Press.

intrinsic growth constraints of 'Technotronic' civilizations.[44] In general, even when the limitations on global growth are accepted as a given, they are attributed to such allegedly transient factors as environmental problems, resource shortages, market fluctuations, deficient planning strategies, or the side effects of 'bad' technologies.

However, despite prevailing technological optimism, the expectations of "unlimited production for all" increasingly appear highly speculative. As the Western ideal of directional progress is closely aligned with the expectations of *unbounded* accumulation of capital and *infinite* technological growth, the *limited potential* of scientific and technological advancements would undermine this historical model in its conservative, liberal, and socialist varieties. As we will see in the next chapter, already in the early nineteenth century, Schelling denounced the idea of infinite progress as an empty fetish of the time.

That is why, apart from problems of political, economic, or cultural tensions, growth constraints have to be raised as a question of the limited growth potential of any technology-based socioeconomic system, no matter how advanced or complex it may be. Such analysis could be undertaken only within an autonomous methodological framework of the philosophy of technology, where technology is not regarded as an epiphenomenon of cultural, environmental, or other factors external to the intrinsic logic of technological progress itself. With humanity fully submerged in a technological environment, the philosophy of technology could be viewed as a form of anthropology, a philosophy of man himself. Nevertheless, as Langdon Winner observed, "At this late date in the development of our industrial/technological civilization, the most obvious observation to be made about the philosophy of technology is that there really isn't one."[45] In the spirit of the age, researchers often substitute the foundational questions with complex computer modeling techniques, leaving their socioeconomic models, at best, methodologically incomplete. As a prominent example, a sophisticated predictive model developed by the Club of Rome advocates of 'zero growth' policies, is based on the assumption of inevitable lag in market response to potentially devastating environmental 'overshoots,' while the issue of intrinsic logic and possible

---

[44] Huntington, S. P. (2011). *The Clash of Civilizations*. Simon & Schuster. Wallerstein, I. (2004). *World-Systems Analysis: An Introduction*. Duke University Press.
[45] Winner, L. (1986). *The Whale and the Reactor*. University of Chicago Press.

limits of qualitative technological progress is left beyond the scope of their study.[46]

A similar problematic emerges in the discussions of the related issue of the infinite progress of science (to be more properly dubbed 'technoscience'). The idea of potential limits to scientific progress has been explored, typically, in the heteronomous frameworks of practical constraints external to the internal structure and dynamic of science itself; among the frequently cited limitations are human psychology, cultural preferences, or socioeconomic problems. When the issue of limits is examined from the allegedly theoretical perspective, the arguments typically proceed within a narrowly defined branch of fundamental science such as thermodynamics or mathematics; for example, Gödel incompleteness theorems is often used by the defenders of infinite scientific progress. Richard Feynman summarizes debates over the prospects of fundamental science in the following way: "This thing [qualitative scientific progress] cannot keep on going so that we are always going to discover more and more new laws… What happens in the future is either all laws become known… or it may happen that the experiments get harder and harder to make, more and more expensive… But I think it has to end one way or another."[47] In the final analysis, as deep as the 'end of science' arguments may look on both sides of the aisle, what they lack is a thorough, meta-scientific, epistemic analysis of scientific progress, its qualitative scope and temporal extent.

As a matter of fact, the persistent resistance to questioning the long-standing ideology of scientific and technological optimism arises from the feeling that it is the essential human nature that is being questioned. Yet, by revealing, and thus transcending, the limits of instrumental reason, we may discover the genuine universality of human existence, satisfying our long-standing longing for a viable 'solution to the riddle of history.' Human freedom should not become a hostage to the idea of infinite economic or technological expansion, falsely conceptualized as 'progress.' In this context, we may need to reassess the viability of "the elimination of scarcity and therewith all major social problems… in finite time" and adjust our long-term social and economic planning to the limited growth

---

[46] The 'limits to growth' and 'end of science' debates are explored in the last chapter. Meadows, D, Randers, J., Meadows, D. (2004). *Limits to Growth: The 30-Year Update*. Chelsea Green Pub. Co.

[47] Feynman, R. (1994). *The Character of Physical Law* (p. 166). The Modern Library.

models, focusing on personal freedom and humane socioeconomic relations "as if people mattered."[48]

With a political power structure rooted in the productivist social mythology, any meaningful changes would involve the gradual *clearing* (in Heidegger's terms) of social consciousness from consumer mentality and awakening from the prejudices of unbounded growth. The real change in historical fortunes is inseparable from breaking the culture of consumerism through the existential grass-root renaissance of man himself, with consumers enabled to define their true wants and establish free and wholesome relation to the world. It is the essential human nature that represents a genuine medium of history, and the only form of universal history is one based on personal existential choices instead of the 'good will' of socially manufactured 'majorities' or the 'expertise' of benevolent elites.

The riddle of human liberty could be illustrated by Dostoevsky's parabola of the Great Inquisitor, where the Inquisitor in Chief proclaims the ideal of freedom as an obstacle to establishing the true kingdom of moral ends. He emphatically asserts that the keys to this kingdom are in the hands of a 'power elite' with the privilege to define the meaning of freedom and happiness and lead the people toward the golden age of an ultimate 'end of history.' The story is an early warning for all who claim an exceptional status for their favorite ideology or a vision of history, independent of whether a self-proclaimed elite is to be represented by the Inquisition, a revolutionary party, or a liberal 'meritocracy.'[49] Paraphrasing Dostoevsky, it is the human soul that is the ultimate battleground for the future of humanity.

---

[48] Gendron, B. (1977), p. 28. Schumacher, E. F. (1999). *Small Is Beautiful: Economics As If People Mattered*. Hartley & Marks.

[49] Dostoevsky, F. M. (1976). *The Brothers Karamazov* In Dostoevsky, F. M. *Collected Works* (Vol. 14, p. 224–239). Science Pub. (in Russian).

CHAPTER 2

The Dialectic of Knowledge and the Idea of History

Enlightenment is the Courage to Use Your Own Reason
—Kant

During the Middle Ages, science and philosophy were judged by their support of theology, and systematic thought of the time proceeded within the tenets of Platonic and Aristotelian traditions. Aristotle, better known in medieval Europe than Plato, was referred to as the Philosopher. The Renaissance marks a transition from a theological to an anthropocentric view of the world, with man replacing God as a moral center of the universe and an engine of historical change. The anthropocentric spirit of the Renaissance is famously articulated in the *Oration on the Dignity of Man* by Giovanni Pico della Mirandola (1463–1494). Though still focused on the union with God and influenced by the mysticism of the Kabbalah, he recognizes man for "the acuteness of his senses, the inquiry of his reason and the light of his intelligence," allowing him to be an "interpreter of nature"; above all, Mirandola emphasizes the dignity of man as a master of his own destiny.[1] The intellectual climate of the Renaissance provided fertile soil for the philosophic spirit of modernity, with its newly found historical mission and the power not only to interpret but also to shape

---

[1] Giovanni Pico della Mirandola. (1956). *Oration on the Dignity of Man* (pp. 3–4). Henry Regnery.

human social and natural environment according to the principles of reason.

The free spirit of the new Age of Reason is prominently expressed by the celebrated French philosopher and scientist René Descartes, widely regarded as the 'father' of modern European philosophy. The Cartesian principle of *cogito*—I think, therefore I exist—affirms the leading role of human reason in the pursuit of knowledge. Moreover, the unbounded expansion of knowledge is seen as a driving force of progress and a principal tool in the amelioration of the human condition. In this spirit, Descartes proclaims that universal science (Scientia Universalis) "can elevate our nature to its highest degree of perfection."[2] However, in the tribute to Scholastic thought, while man is endowed with an autonomy of reasoning, God retains the *ontological* status of a creator and sustainer of *thinking* and *material* substances. As a result, man and God are entangled in a circular interplay of epistemic certainty and ontological actuality. The *Cartesian circle* of knowledge and being, implying the incommensurability of human reason with the actuality of God, entails the infinity of the historical progression toward human mastery over the world.

The Cartesian dialectic of knowledge and being found a new form in Kantian transcendental philosophy. Kant's objective is to account for the human ability to act in compliance with both the *formal* causality of reason and the *effective* causality of nature. Raising rationalism to a new level, he transforms a unitary act of Cartesian *intellectual intuition* into the multi-faceted activity of *pure reason*. In this way, he develops the refined version of Cartesian dualism by radically separating the *formal* architecture of nature (constituted by the creative spontaneity of reason) from the *actuality* of the 'thing in itself.' The dualism between the noumenal core of human experience and its phenomenal shell extends even to man, sharply divided between the realms of moral freedom and material necessity. Due to the chasm between the empirical and transcendental aspects of human nature, historical progress toward the moral 'kingdom of ends' is conceived as an *infinite* approximation of human material existence to the categorical requirements of practical reason.[3]

Trying to reconcile the dualism of moral normativity and natural causality, Kant's celebrated follower, Johann Gottlieb Fichte, develops a

---

[2] Descartes, R. *Qeuvres de Descartes publees par Charles Adam et Paul Tannery* (Vol. 1, p. 339).

[3] Kant, I. (2012). *The Critique of Practical Reason* (Pt. 2, chap. 2 (IV)). Dover Pub.

holistic conception of reality as a product of dialectical self-realization of the absolute I. In the spirit of the age, he understands philosophy as a universal *science of knowledge,* allowing "to *order all their* [human] *relations with* FREEDOM *according to* REASON."[4] Elevating subjective activity to the level of universal creative power, Fichte construes history as a unitary process driven by the moral normativity of the I and, in the progressive spirit of the Enlightenment, views science as an instrument of historical progress and social amelioration. At the same time, in his system, nature is effectively reduced to hardly more than a historical epiphenomenon of the dialectical unfolding of the I. In effect, in his philosophy of freedom, ideal human activity is entirely detached from the material conditions of human life.

In Hegel's comprehensive synthesis, the subjective dialectic of Fichtean I is transformed into the circular dialectic of the Absolute (God). On this background, Hegel develops an essentially close-ended conception of history, in which the Absolute finds its objective realization in the *concrete universal* of the State. The intrinsic logic of the historical dialectic of matter and spirit is succinctly expressed by the motto: *what is real is rational, and what is rational is real.*[5] At the same time, contrary to the ideals of the Enlightenment, man is viewed by Hegel not as a driving force of progress but as a passive witness to the historical unfolding of the Absolute, with human freedom dissolved in the 'general will' of the State. This historical conservatism and the radical panlogism of the Hegelian system signify an eclipse of German transcendental tradition.

Hegel's historically influential left-wing followers set the objective to free philosophy from the tenets of objective idealism and show that human activity is a genuine driving force of social progress. This movement culminates in Marx's dialectic of labor, with philosophy conceived as the *revolutionary praxis* changing the world.[6] In Marx's ontology of labor, human activity becomes a 'measure of all things, existing as they exist and non-existing as they do not exist.' In the following sections, our principal objective is to demonstrate how the scholastic problematic of *formality*

---

[4] Fichte, J. G. (1977). *Characteristics of the Present Age* (D. N. Robinson, Ed.) (p. 5). University Pub. of America.

[5] For the dialectic of abstract and concrete, as a 'secret' of Hegel's philosophy, see Sterling, J. H. (2017). *The Secret of Hegel.* Forgotten Books (Reprint of 1898, 2nd ed.). This approach is reiterated in Marx's statement that 'reason has always existed but not always in a reasonable form.'

[6] Marx, K. "Theses on Feuerbach" (thesis 11).

and *actuality* is recapitulated in the *Cartesian circle* of knowledge and being, the Kantian dualism of formal causality of reason vs. material causality of nature and the Hegelian dialectic of the Notion. In the next two chapters, we examine how the Cartesian epistemological circle is carried over to Marx's conception of labor as a unity of universal and material modes of human activity.

IN THE CIRCLE OF KNOWLEDGE AND REALITY

At the heart of Cartesian rationalism is a celebrated *method of doubt* focused on the rational reconstruction of human experience. The first and the most fundamental outcome of radical doubt is a unity between epistemological certainty and its ontological import: "It is not possible for us to doubt that we exist while we are doubting and it is the first thing we come to know when we philosophize in an orderly way."[7] Hence, the proposition 'I think, therefore I exist'—*cogito ergo sum*—becomes the first and the most evident postulate of the Cartesian method. The Cartesian cogito is intended to 'bracket' (phenomenologically speaking) all empirical modifications of the ego, including its psychological capacities and reveal its universality as a subject of thought.[8] Descartes reiterates this idea, stating that "I is a substance, whose whole essence or nature is simply to think, and which does not require any place, or depend on any material thing, in order to exist."[9] The thrust of his argument is on the unity of knowledge and being, realized through the synthesis of the intuitive act of *cogito* and its objective content.

---

[7] Descartes, R. *Principles of Philosophy* (Pt. I, par. 7) In Descartes, R. *Selected Philosophical Writings* (J. Cottingham, R. Stoothoff, D. Murdoch. Eds.). Cambridge University Press.

[8] The method of doubt gave rise to multiple interpretations, some even questioning Descartes' rationalism. For example, Rockmore remarks that "it is methodologically unsound to accord priority to epistemology prior to a view of the subject" and argues that it is "incorrect to separate the logical conditions of knowledge from man's psychological capacities because when this is done, there is no assurance that a subject adequate for the epistemology in question in fact exists." However, the separation of thinking substance from its material (psychological) alter ego is at the core of the Cartesian method, intended to bridge the gap between knowledge and reality, the universality of man and the material causality of nature. Rockmore, T. (1980). *Fichte, Marx, and the German Philosophical Tradition* (p.159). Southern Illinois University Press.

[9] Descartes, R. (1988). *Discourse on Method* (IV, pp. 33–34) In Descartes. *Selected Philosophical Writings* (J. Cottingham, R. Stoothoff, D. Murdoch. Eds.). Cambridge University Press.

A transition from the epistemological certainty of 'I think' to the existential import of 'I exist' is justified by the appeal to God. While clear and distinct ideas are being *constructed* by the mind, their existential import is based on the *actuality* of God: the power to create himself and sustain his existence is not present in man but requires the omnipotent God. In the third of his *Meditations,* Descartes underlines that certainty of the *cogito* is not a guarantee of his sustained existence as a thinking substance: "I observe that there is no such power…through which I can bring it about that I myself, who now am, will also exist a little later …from this fact I know most evidently that I depend upon a being other than myself." He explains that in the idea of God as a perfect being, man finds not "merely the possible and contingent existence… but utterly eternal and necessary existence… Now, on the basis of its perception that necessary and eternal existence is contained in the idea of a supremely perfect being, the mind must clearly conclude that the supreme being does exist."[10] Descartes reaffirms the same idea in the "Arguments… Drawn up in Geometrical Fashion": "But necessary existence is contained in the *concept* of God. Hence, it is true to affirm that necessary existence *exists* in Him, *or* that God Himself exists."[11]

At the core of this argument is an *ontological proof* of the existence of God, introduced into the theological discourse by St. Anselm. However, Descartes' reasoning is closer to the more refined version of the proof offered by Duns Scott, who explicitly emphasizes the *necessary* character of God's existence, a point Descartes considers essential. Elaborating on this idea, in the third of his *Meditations,* he assigns to God not just "potential" but "actual or formal" existence. The alleged identity between the *actual* and *formal* existence draws Descartes closer to scholastic realism: only a being that exists 'formally' could guarantee "the objectivity of the idea," that is, its correspondence to reality. The Cartesian identity between formality and actuality becomes a central principle of Hegelian panlogism.

Despite its scholastic roots, the distinctive feature of the Cartesian approach, signifying a break from the scholastic tradition, is that God's existence is inseparable from the power of human activity, the *cogito*. In

---

[10] Descartes, R. Meditations (Pt. 3). Descartes, R. *Principles of Philosophy* (Pt. I, par 14).

[11] Descartes, R. (1985). "Arguments Demonstrating the Existence of God and the Distinction Between Soul and Body, Drawn Up in Geometrical Fashion" In *The Philosophical Writings of Descartes* (J. Cottingham, R. Stoothoff, D. Murdoch. Trans.). (Vol. 3, prop. 1). Cambridge University Press.

"Reply to Objections 2," he emphasizes the reciprocal relationship between the ability of the human mind to construct an idea of God and the actuality of God's existence: "**the whole force of my argument lies in the fact that the capacity for constructing such an idea** [of God] **could not exist in me, unless I were created by God.**" In the idea of God, the subject and God join as the two complementary aspects of unitary human activity; their dynamic interplay allows the act of *cogito* to transcend its *epistemic* origin and reach the *ontological* base represented by the Creator. In this context, Descartes insists that *cogito ergo sum* is not a syllogism because a syllogism would require a premise "Everything that thinks exists," which does not have an intuitive certainty of the 'I think.' From this perspective, the Cartesian argument should be considered not as a formal deduction but as a constructive ideation, touching the Absolute in its *in itself* existence.

Despite its apparent novelty, the circular dialectic of man and God, knowledge and existence, can be traced to St. Augustine's *Confessions*, where he emphatically asks how one could believe in God without knowing him and know God without believing in him in the first place.[12] Augustine resolves this puzzling circularity by appealing to religious *faith*. By contrast, Descartes' aspiration is to find a firm foundation for *knowledge*—including the subjective certainty of God's existence—in reason alone. This difference underlines the Cartesian breakup with the scholasticism of the preceding era and the decisive transition to a new, anthropocentric ideological paradigm of the Age of Reason. In this context, Koyre refers to Cartesian circle of knowledge and being as "necessary and natural" and "characterizes not only Cartesian theory but also any epistemology because any epistemology is intended to discover both the objective sense and the [subjective] means of knowledge. It is contained in any act in which knowledge has itself as an object. In this self-reference, knowledge touches the Absolute."[13] The intrinsic circularity of the Cartesian

---

[12] Augustine. (2008). *Confessions* (H. Chadwick. Trans.) (Book 1, ch. 1). Oxford University Press.

[13] Koyrè, A. (1923). *Descartes und die Scholastic* (s. 97). Bonn. Dostoevsky, in the *Demons*, his last major work conceived as a prophetic indictment of the Russian revolutionary movement, presents the Man-God dilemma with full clarity. One of the 'demons,' representing the Westernized Russian socialists, argues: "If there is God, then he is omnipotent; if there is no God—I am the omnipotent God." In a spirit of Cartesian *cogito*, the judgment on the existence or non-existence of God is construed as a free transcendental act of self-positing that entails either the existence of God (in which case the actuality of ego is grounded in

argument bears the seeds of the "whole future dialectic of dependence and independence of man, the dialectic of Man-God and God-Man" that repeatedly reemerges in European intellectual tradition.[14]

The reciprocal relationship between knowledge and being is replicated in Descartes' philosophy of nature. He argues that the idea of thinking substance could be only constructed on the background of the opposite, but equally clear and distinct, idea of material substance. The ontological reality (*actuality*) of both substances is guaranteed by their *formal* representation in God as a creator and sustainer of the universe.

Unifying epistemology and natural philosophy, Descartes construes mathematics as a model of rational thought and views science as a principal means for the advancement of human knowledge. One of the leading mathematicians of his time and a founder of analytic geometry, he envisions the project of universal science based on the development of universal mathematics (Mathesis Universalis) and conceives nature, in its 'clear and distinct' form, as a *mathematical manifold,* with extension as its *primary quality*. However, the limitations of Descartes' rationalism do not allow him to productively reconcile the *formal* invariability of natural laws with *material* causality. Despite the progressive appeal to man as a rational agent and a "master and possessor of nature," Cartesian dualism comes into conflict with an emerging, in the sixteenth-seventeenth century, new scientific paradigm associated with such names as Copernicus, Kepler, Tycho Brahe, Galileo and Newton. For Newtonian physics, scientific explanation is tightly linked to observation and inductive generalization rather than rooted in a pure rational deduction from the a priori ideas.

On the philosophical front, Descartes' rigorous rationalism was attacked by David Hume (1711–1776), who distinguishes between the "matters of ideas" and the "matters of fact," arguing that all *ideas* are derived from the inward or outward human 'sentiments' ('impressions' in his terminology) and the concept of their *necessary connection* is just a product of repetition

---

God's existence—this is Descartes' position) or (if God does not exist) one's existence as Man-God—this is Dostoevsky's *existentialist inversion* of the Cartesian doubt. The suicide of Kiriloff, faced with this dilemma, is to symbolize a destructive character of the autonomy of reason separated from the national soul. Достоевский Ф. М. (1974). *Бесы* [Demons] In Достоевский Ф. М. *Полное Собрание Сочинений* [Collected Works] (Том 10, стр. 470). Изд. Наука [Science Pub.].

[14] Visheslavtsev, B. P. (1994). *Etika Preobrazennogo Erosa* (1931) [Ethics of the Transformed Eros] (p. 138). Moskva: Izdatelstvo Respublika. A noted philosopher, he was expelled from Russia in 1922, together with over 200 other prominent intellectuals.

and habit (i.e., more or less justified, inductive generalization); by extension, this assessment applies to the notion of *causation* "discoverable not by reason but by experience." Such critique pertains to the core of both Cartesian epistemology and the philosophy of nature. The productive conundrum of reason and nature freedom and necessity finds a new expression in Kantian criticism and makes a significant impact on the consequent development of European philosophy, including Marx's anthropocentric ontology of labor as both universal and material activity.

Addressing the problematic of Cartesian dualism, Baruch Spinoza (1632–1677) transformed it into a monism of one infinite substance conceived as an undifferentiated totality of attributes. While for Descartes, God is a creator and sustainer of thinking and material substances, for Spinoza, God is an *identity* of both. At the same time, an ambiguous relationship between the order of nature and the order of ideas opens the way to a broad range of interpretations of Spinoza's system, from *pantheism* to straightforward *materialism* (by the Marxists) and *dogmatism* (by the followers of Kant). For the consequent development of European rationalism, the heuristic imperative is to conceive this identity as, using Hegel's expression, the substance "acting within itself" and, thus, revealing its internal *dynamic* as a unity of spirit and nature, moral autonomy and natural necessity. Only then (critical) philosophy could demonstrate how man can act as both a free rational agent and a material being.

An important step toward unifying the activity of reason with the passivity of nature was made by G. W. Leibniz (1646–1716), a celebrated representative of European rationalism and a preeminent mathematician. Refining Descartes' 'geometrical reasoning' with the infinitesimal calculus, he applies the principle of the mathematical *continuum* to the philosophy of nature and metaphysics and introduces a paradigmatic idea that activity is an essential characteristic of a substance. In his system, the world comprises the *infinite* hierarchy of self-reflecting and eternally active entities—*monads*—'participating' in the "whole of the universe."[15] Foreshadowing the Kantian conception of *transcendental* synthesis, Leibniz emphasizes that the mind is "not merely the scene of ideas but

---

[15] Leibniz (1898). *Monadology*. In *Leibniz: The Monadology and Other Philosophical Writings* (Robert Latta. Trans.). (Prop. 66). Oxford University Press.

their source and origin" and famously remarks (in the polemic with Locke) that there is nothing innate in the mind except the mind itself.[16]

For both Spinoza and Leibniz (as for the Enlightenment, in general), human history and moral values are at the very core of metaphysics. Notably, the title of Spinoza's magnus opus is *Ethics*. For Spinoza, the ultimate goal for man is life "under the entire guidance of reason" in a democratic society.[17] He emphasizes that the objective of scientific progress is "to direct all sciences to one end and aim, so that they may attain to the supreme human perfection," adding that "whatsoever in the sciences that does not serve to promote our object will have to be rejected as useless."[18] Such a position represents a departure from the predominantly pragmatic attitude toward science as an instrument of mastery over nature, characteristic of as diverse 'fathers' of modern philosophy as Descartes and Francis Bacon.

What is at stake is the nature of human civilization and history. While Cartesian rationalism implies that scientific and technological advancements are leading forces of historical progress, the alternative view emphasizes the disparity between technological and scientific vs. moral perfection. The latter position was conspicuously expressed by Jean-Jacques Rousseau (1712–1778), who considered science and technology a decadent power corrupting essential human nature. As he says in the *Discourse on the Arts and Sciences* (1750), "Our souls are corrupted as our science and arts advance to perfection." The recurrent tension between the strive for moral values and mastery over nature is replicated in Marx's dichotomy between the goal of emancipating humanity from the sense of having vs. his adamant reliance on infinite technological expansion as a foundation of open-ended historical progress.

In Leibniz's picture of the universe, the present contains a blueprint of the future, conceived as a "complete harmony" of the "moral kingdom of grace" and the "material kingdom of nature." This harmony is to be accomplished as an infinite progression toward the universal monarchy

---

[16] Leibniz, G. W. (1996). *New Essays on Human Understanding*. Introduction. Cambridge University Press (Written in 1704, published in 1765). Cassirer, E. (1951). *The Philosophy of the Enlightenment* (p. 121). Beacon Press.

[17] Spinoza, B. (2001). *Theologico-Political Treatise* (Samuel Shirley. Trans.). (Prop. 16). Hackett Pub.

[18] Spinoza, "On the Improvement of the Understanding." In Spinoza, B. (n/d). *Improvement of the Understanding, Ethics and Correspondence* (R. H. M. Elwes. Trans). Beling Tetens Publisher.

(Monarchie Universelle) of "the moral world within the natural world."[19] In this way, the moral teleology of final causes finds its realization in the material realm of efficient causes. The pre-established harmony of the universe is a product of God, who created the *best possible world* and directs it toward the ultimate perfection of the kingdom of grace (Voltaire, with his usual wit, ridicules this idea in *Candide*).[20] Due to the chasm between the material universe and the perfection of its Creator, this aim remains an ideal goal of the *infinitely* unfolding historical progress (strictly speaking, this conception of progress applies more to the universe as a whole than to human history per se). Leibniz's historical optimism and the idea of unity between the "kingdom of grace" and "kingdom of nature" resonate with Kant's idea of the moral kingdom of ends and the "secret plan of nature," while his analysis of the activity of reason foreshadows Kantian transcendental methodology.

Kant's criticism brought into focus a wide range of problems in metaphysics, the philosophy of nature, the philosophy of history and ethics, and opened a new era in the development of European philosophy. In particular, the problematic of the Cartesian interplay of spirit and nature provided a fertile ground for Kant's transcendental critique of reason, leading to the Hegelian close-ended dialectic of the notion and its 'inversion' in Marx's dialectic of labor.[21]

---

[19] Leibniz. *Monadology* (Props. 86, 87, 88). The interaction between the spirit and the body is also addressed in *Principes de la Nature et de la Grace Fondees en Raison* in *Die Philosophischen Schriften Von Gottfried Wilhelm Leibniz* (1875–1890) (Bd. VI, pp. 598–606).

[20] The theological ambiguity of Leibniz's philosophy is represented at different levels. In particular, it is questionable whether "Leibniz's logical and metaphysical premises... compatible with his acceptance in *Theodicy* of the ideas of sin and eternal sanctions." Copleston, F. S. J. (1963). *A History of Philosophy* (Vol. IV, p. 334). Image Books.

[21] On this background it is hardly justifiable to claim, as Rockmore does, that "Kant's epistemological stance is relatively anti-Cartesian since... if man is merely a spectator, then knowledge is impossible." If Kant's stance is "relatively" anti-Cartesian, it is not because, for Descartes, man is only a "spectator" but because Kant further sharpens Cartesian anti-psychologism to develop his a priori dialectic of the *cogito*. In a trivial sense, Kant is as anti-Cartesian, as Fichte is anti-Kantian or Hegel is anti-Fichtean. Rockmore, T. (1980). *Fichte, Marx, and the German Philosophical Tradition* (p.159). Southern Illinois University Press.

The Limits of Reason and Dualism of History

Confronted with the problems of Cartesian rationalism, Kant says that it was Hume who awakened him from the "dogmatic slumber."[22] He explains that Hume "demonstrated irrefutably that it was entirely impossible for reason to think *a priori* and by means of concepts such a combination as involves necessity. We cannot at all see why, in consequence of the existence of one thing, another must necessarily exist, or how the concept of such a combination can arise *a priori*."[23] For Hume, all *a posteriori* (empirical) knowledge is contingent, while all non-contingent (necessary) knowledge is analytic (not adding factual information). The difficulties in reconciling analytic and synthetic knowledge and the controversy around the notion of necessity expose, in full force, a sharp split between the rationalist and empiricist pictures of the world.[24]

In the concession to empiricism, Kant rejects the existential import of the Cartesian *cogito* along with the ontological proof. The celebrated Kantian objection to the *cogito* argument is that existence is not a predicate. In fact, Kant makes a more subtle point—he distinguishes between *being* and *existence*. He says that only if we consider the "thing in itself" then such a being (Sein) means the same as existence (Dasein)."[25] Schelling explains Kant's critique in the following way:

> Thinking is a particular *sort of being*, even more, 'cogitans' means—I am in a **state** of thinking," because I think or doubt only in the *act* of doubt. "Therefore, 'sum,' contained in 'cogitans,' means only 'sum qua cogitans'— 'exist as thinking,' i.e. in this particular mode of existence which is called thinking[26]

---

[22] Kant, I. (1966). *Prolegomena to Any Future Metaphysics* (Lucas, G. Trans.) (p. 9). Manchester University Press.

[23] Kant, I. (1966), p. 6.

[24] Hume does not employ such terminology; he distinguishes between (analytic) *relations of ideas* and (synthetic) *matters of fact*; for a more detailed discussion, see Copleston, F. (1959). *A History of Philosophy*, vol 5, pt. 2, chap. 14.

[25] Kant, I. (1979). "The one possible basis for a demonstration of the existence of God" (1, 1, 2, trans. modified). Abaris Books. Here we deal with Kant before Hume "interrupted [his] dogmatic slumber;" nevertheless, the understanding of existence as the "absolute positing of the thing," different from the mere being, constitutes a basis for the conception "thing in itself" central to his criticism.

[26] Schelling, F. W. J. (1975). *Zur Geschichte Der Noueren Philosophie*. (s. 14). Darmstadt: Wissenschaftliche Buchgesellschaft.

In this interpretation, the 'thinking thing' is not a substance that *exists* (in itself) but a particular being that *subsists*; thinking involves only a particular *mode* of existence but does not constitute existence per se. Explicating further Kant's critique, Schelling explains that 'I think' or 'I doubt' presupposes something (mental) in which this thought or doubt is reflected: "only as soon as this **second** something accepts the first one to be identical to itself, only then I say—'I think.'"[27] Therefore, 'I think' cannot be considered as a direct act of intellectual intuition (grasping an object of thought in its ideal purity) but as a *synthetic activity* (of *apperception*): I cannot be aware of something without myself being aware of being aware, etc. In such a regression of the synthetic mental acts, their ultimate unity of 'I think' represents an *invariant* structure of mental experience rather than 'touches' the ontological reality of either the thinking substance or God.

Kant employs similar reasoning in his analysis of Descartes' ontological proof of God's existence. While Kant's critique is valid, it is not quite precise. He construes Descartes' argument as if it states: I find in myself the idea of the perfect being, but the idea of perfection contains in itself existence; therefore, God exists. Kant objects: existence is not a predicate and, therefore, it does not belong to the idea of perfection or omnipotence. However, Descartes does not precisely say what Kant makes him say. He makes a more subtle distinction between contingent and necessary modes of existence. Only because the idea of a perfect being contains eternity and necessity does God exist. However, what Descartes, in fact, proves is that *if* God existed, he would exist *necessarily*, but this does not mean that God *necessarily exists*. A premise of the Cartesian argument is about the *way* in which God exists as 'completely eternal and necessary,' while the conclusion is about the existence as such.[28] Schelling finds here a logical fallacy: *plus in conclusion est quam fuerat in premisses* (more in conclusion than in the premises).[29]

To generalize this criticism, what exists is not I as a substance or God as an absolute creative power but the *formal act* in which God and I emerge as two inseparable logical moments of the (Cartesian) epistemological

---

[27] Schelling, (1975), s. 15.

[28] Leibniz, in his remarks on Descartes' *Principles of Philosophy*, emphasizes that the necessity of God's existence must be proved first to conclude that God exists. Leibniz, G. W. (1697). *Animadvesiones in Partem Generalem Principiorum Cartesianorum*.

[29] Schelling, F. W. J. (1975), s. 18–19.

circle. The clear and distinct character of the *cogito* does not establish the existence of either the thinking substance or God. What it does establish is a certain (relational in nature) *invariance of subjective experience*, in which the idea of my existence is inseparable from the idea of its assumed ontological base (God). The problem of Descartes' method is the implicit identity between the *universality* of clear and distinct ideas and their *existential import*, rooted in the scholastic contention that *formal completeness* entails *ontological reality* (actuality). The connection (and tension) between formal and material causality becomes a powerful heuristic idea for the consequent development of transcendental philosophy and constitutes a dominant theme of Hegelian dialectic and its consequent 'inversion' by Marx.

Despite his critique of Descartes, Kant concurs that spontaneous (a priori) constructive activity of the mind does, after all, have an existential import in one, though vital for human knowledge, field of mathematics. Unlike Hume, he maintains that the necessity of mathematical propositions does not represent the 'relation of ideas' based on symbolic conventions but rather the 'matters of fact'; therefore, these propositions are not *analytic* but *synthetic*.[30] Accordingly, the guiding idea of his *transcendental analysis* of reason is to reveal the conditions under which the 'synthetic a priori' knowledge is possible in mathematics and investigate whether and how similar (factual) synthesis is possible in natural science and philosophy.

Addressing this issue in *The Critique of Pure Reason*, Kant explains that human reason acquires knowledge only of what it produces after a plan of its own and claims that mathematics is the only science that fits the bill because "mathematical knowledge is that which reason gains from the **construction** of concepts. And to **construct** a concept is to demonstrate **a priori** the intuition corresponding to it." He emphasizes the creative nature of the *constructive activity*: "In the empirical intuition [of a geometrical figure] we consider always the *act* of the construction of the

---

[30] For Hume, mathematical judgments are analytic and represent relations of what he calls 'impressions.' Leibniz also considers mathematical judgments as analytic (because they can be derived from the laws of contradiction and identity); but his position could also be interpreted in a spirit of (mathematical) constructivism; for constructivist elements in Leibniz's philosophy of mathematics, see Cassirer, E. (1902). *Leibniz's System* In *Seinen Wissenschaftlichen Grundlagen*. Marburg. Kant sees mathematical knowledge as both *necessary* and *synthetic*.

concept..." that allows to present the products of the a priori construction "in concreto."[31]

The activity of mathematical construction involves two interrelated steps: the (a priori) act of transcendental *construction* and the consequent *objectification* and 'synthetic assembly' of its products through the *a priori forms of sensibility*—space and time. The activity of reason justifies the universality and necessity of mathematical knowledge, while the medium of space and time yields its synthetic character. As an example, geometry requires pure intuition of space and arithmetic—of time; in line with the Cartesian mechanical picture of the universe, Kant includes into this group also 'pure mechanics.' The mechanism of construction and objectification becomes a foundation of dialectical apparatus developed by Kant's immediate followers, Fichte and Hegel, and makes an impact on Marx's philosophy of technology and the theory of organ projection developed by Ernst Kapp and Ludwig Noire (explored in Chap. 4).

Kantian methodology can be presented as a transcendental 'inversion' of the famous Plato's Myth of the Cave, with a subjective activity of productive imagination playing the role of the sun and the a priori forms of sensibility, representing the screen on which the 'shadows' of reality are being assembled and processed by pure reason. The products of such an a priori activity constitute what appears to us as (natural) 'objects,' while the reality *in itself* remains beyond the grasp of human knowledge.

Applying this methodology to natural science, Kant concedes (to empiricists) that we have no idea about the causality of nature as it is *in itself* but maintains that we still can develop a rational picture of the world as it appears *for us*. Unlike mathematics, where a priori synthesis is based on the spontaneous activity of productive imagination, natural science requires the external stimuli of sense data (from the thing in itself), consequently 'assembled' and presented on the a priori 'screen' of space and time. As a result, an object of knowledge emerges as a product of sophisticated synthesis involving the a priori forms of sensibility, the a priori forms of understanding and the a posteriori (in origin) sense data. From

---

[31] Kant, I. *The Critique of Pure Reason* (Max Muller. Trans.) (B, 739). Kant, I. *Prolegomena to any Future Metaphysics* (1, 7). Such an understanding of a synthetic a priori judgment foreshadows the twentieth-century mathematical constructivism, including intuitionism of Brouwer; see Weyl, H. (1949). *Philosophy of Mathematics and Natural Science*. Princeton University Press. On the notions of *analytic* and a priori, see also H. Putnam (1983). "Analyticity and APriority: beyond Wittgenstein and Quine" In *Realism and Reason. Philosophical Papers* (Vol. 3). Cambridge University Press.

this perspective, a fundamental notion of objectivity in general and 'physical object' in particular represents a complex product of the a priori construction. In effect, Kant turns the invariability of the laws of nature into the invariability of the laws of reason and, considering that mathematics is the only genuinely a priori science, nature, as we know it, is effectively reduced to a mathematical manifold. In line with the Cartesian ideal of mathematical physics, Kant claims that natural science contains as much 'science' as it contains mathematics.

The wall between reality *in itself* and its 'objective' appearances applies even to the (transcendental) subject who has access to himself (as a noumenon) only through his objectified representation on the 'screen' of a priori forms of sensibility; in other words, pure subjectivity, in its immanency, remains *transcendent* to its phenomenal alter ego in the mundane world of sense data. Here, the immanent *act* of internal construction and its phenomenal *product* appear as the two complementary aspects of *transcendental* activity.

The dual character of a priori synthesis can be illustrated by Bohr's principle of complementarity and (in this context) his interpretation of Heisenberg's uncertainty principle (we can accurately detect either the speed of an elementary particle or its position, but not both).[32] The uncertainty in question relates to the inseparable unity of the (macro) measurement procedures and the mode of objectivity assigned to corresponding quantum events. In line with Bohr's interpretation, the subject and the object are the two complementary aspects of what a physicist may qualify as objectivity of physical and, especially, quantum events. Likewise, in the 'uncertainty' of transcendental activity, we either intuit the act of construction itself or capture its objective representation in the phenomenal world. The holistic act of knowledge is constituted by the progressively unfolding oscillation of consciousness between these two complementary modes of activity. An outcome of this convergent series of mental states is what is perceived as the 'objectivity' of human experience dubbed as 'objective reality.' This phenomenology of constructive activity and its objectified products becomes a guiding principle of Fichte's attempt to overcome Kantian dualism in his version of transcendental methodology.

The complicated balancing act between the a priori forms of understanding, the a priori forms of sensibility, and the 'thing in itself' is

---

[32] Bohr, N. (1934). *Atomic Theory and the Description of Nature*. Cambridge University Press.

intended to maintain a middle ground between rationalism and empiricism. However, by not allowing theoretical reason to penetrate the inner core of reality, Kant allegedly demonstrated the principal limitations of a *theoretical* attitude to the world. Emphasizing this point, Schelling remarks that "theoretical reason **necessarily** seeks what is not conditioned; having formed the **idea** of unconditioned, and, as **theoretical** reason, being unable to realize the unconditioned, it therefore **demands** the act through which it ought to be realized."[33] What Shelling means is that the idea of the "unconditioned" *thing-in-itself* points to the limits of theoretical reason and, in this way, demands a transcendence of the *theoretical attitude* of pure reason into the *practical attitude* of action. "*The Critique of Pure Reason* first proved," says Schelling, "that no system... can be an object of *knowledge* but can be only an object of practically necessary but *infinite* activity."[34] In other words, Kant's critique entails that any attempt to develop an all-embracing *theoretical* picture of the world would *inevitably* be either *inconsisten*t or *incomplete*. Addressing the epistemic core of this conundrum, F. H. Jacobi (1743–1819) says that we cannot enter Kant's system *without* the thing in itself and cannot remain within it *with* the thing in itself.

The limitations of theoretical reason, revealed by Kant's critique, lead to the standpoint of *freedom* as a key constitutive principle of philosophy, marking a transition from the contemplative conception of philosophy as a *theory* explaining the world to the program of moral and social *praxis*. This approach, conspicuous in Fichte's interpretation of transcendental methodology as a philosophy of freedom, leads to the conception of philosophy as social criticism advanced by the left-wing Hegelians and as revolutionary praxis—by Marx.

It was Kant himself who made a vital step in this direction by introducing the notion of *practical reason*. In the same way that the theoretical reason explains causality, the practical reason provides a basis (*ratio essendi*) for moral law (guiding our behavior in the empirical realm of necessity). In turn, the autonomy of moral law reveals freedom as an essential characteristic of men as rational beings (*ratio cognoscendi*). As in the case of

---

[33] Schelling, F. W. J. (1980). *Philosophical Letters on Dogmatism and Criticism* (Letter 4). In Schelling, F. W. J. *The Unconditional in Human Knowledge: Four Early Essays*, 1794–1796 (Marti, F. Trans.) (p. 167). Bucknell University Press.

[34] Schelling, F. W. J. *Philosophical Letters on Dogmatism and Criticism* (Letter 5 & 6, p. 171).

theoretical reason, the dialectic of practical reason involves the two complementary aspects of a unitary experience: the free transcendental activity and its objectified form of the moral law (we impose on ourselves). This dualism poses a complex problem of distinguishing between *arbitrary will*, guided by the phenomenal realm of everyday experience, and *good will*, guided by the normativity of pure reason.

The harmony between the two complementary levels of human existence—freedom (morality) and material necessity—is secured by the "secret plant of nature," implemented in the historical progression of humanity toward the ideal kingdom of ends. While at the noumenal level of reality *in itself*, moral freedom and natural necessity are indistinguishable, in human experience the harmony of the two worlds can be accomplished only in the infinite progression of humanity as a *species* due to the disparity between the moral telos and the actuality of the historical process. Paraphrasing Bernstein, in the transcendental conception of history, the infinite movement of *the human species* toward the moral kingdom of ends is everything, while the final goal is nothing.

Overall, Kant's Copernican revolution transformed the old-fashioned metaphysics into philosophical anthropology, focused on man, the scope of his powers, and his place in the universe. The divide between the noumena and phenomena, theory and practice, moral law and nature finds its completion in the historical unfolding of human freedom. As a result, history is no longer understood as an instrument of the divine revelation or an aggregate of empirical events but acquires its telos in the depth of human subjectivity.[35] Then, it becomes clear why Kant considers the notion of freedom indispensable for the architecture of pure reason. In Schelling's words, the spirit of critical philosophy is to "strive for immutable selfhood, unconditional freedom, unlimited activity," opening the way to the synthesis of moral teleology and the iron necessity of nature. This ambitious program inspires the Fichtean *philosophy of freedom* and its consequent development by Schelling and Hegel, culminating in Marx's historical *dialectic of labor*.[36]

---

[35] Kant, I. *The Critique of Practical Reason* (Preface).
[36] Schelling, F. W. J. (1980), letters 6 & 9, p. 192.

History as a Triumph of Freedom

Fichte, Kant's follower and junior colleague, claims that his *Science of Knowledge* revealed the true spirit of Kantian criticism by eliminating the chasm between theory and practice, placing practical reason at the foundation of the whole architecture of knowledge. In this way, knowledge allegedly transcends the theoretical limits of epistemic certainty and becomes an instrument of practical volition and "*action itself.*"[37] Emphasizing the practical spirit of his philosophy, Fichte proclaims it "nothing but the analysis of the notion of freedom."

The pivotal step in the transition from the Kantian dualism of freedom and necessity to the Fichtean standpoint of freedom is a reinterpretation of the act of *construction*. While for Kant, the prime examples of genuinely constructive, and thus *rational*, activity are provided by mathematics, Fichte and his junior colleague Schelling maintain that construction plays an even more fundamental role in philosophy. Schelling tackles the problem at its core: "The mathematician, furthermore, is never concerned directly with intuition (the act of construction) itself, but only with the result of the constructive activity." By contrast, "the philosopher looks solely to **the act of construction itself**, which is absolutely internal by its character," that is, "the objects of transcendental philosophy exist only as far as they are freely produced," and the philosopher has direct access to the mechanism of construction (something Kant does not accept). This is possible under two conditions: "First, that one be engaged in a constant inner activity… Second, that one be constantly reflecting upon this activity; in a word, that one always remains, at the same time, both the intuited (the producer) and the intuitant."[38] It is the unity of the first and second elements of constructive activity that constitutes the faculty and "art" of the a priori synthesis. Fichte elaborates on this idea in the following way:

> The office of the synthesizing faculty is to unite the opposites, to think them as one… Now this it cannot do; yet the requirement is there; and hence there arises a conflict between the incapacity and the demand. The mind lingers in this conflict and wavers between the two—wavers between the

---

[37] Fichte, J. G. (1910). *The Vocation of Man* (Book I, Doubt). Open Court Pub. One of the notable influences on Fichte is Salomon Maimon's critique of Kant. Maimon, S. (2010). *Essay on Transcendental Philosophy*. Continuum.

[38] Schelling, F. W. J. (1978). *System of Transcendental Idealism* (Peter Heath. Trans.). (p. 13, trans. modified). University Press of Virginia.

requirement and the impossibility of carrying it out. And in this condition, but only therein, it lays hold on both at once, or, what comes to the same thing, makes them such that they can simultaneously be grasped and held firm; in touching them, and being repulsed, and touching them again, it gives them, in the **relation to itself**, a certain content and a certain extension (which will reveal itself in due course as a manifold in time and space). This condition is called the state of **intuition** (Anschauen). The power active therein has already been denominated earlier the productive imagination.[39]

From the 'producer' perspective, the I, in its "state of intuition," renders human experience as a complex *hierarchy* of the creative acts of intuition (productive imagination) in their "relation to themselves." From the "intuitant" (spectator) perspective, this hierarchy emerges as a "certain [objectified] content" revealing itself (in space and time) as *non-I*—the world. A reemerging conflict between the creative power of the mind and its objectified content is being continuously resolved (negated), providing a foundation for an expanding synthetic activity of the I. In this process, the distinction between the *a priori* and *a posteriori* loses its absolute character and can be defined only in the context of the (relative) 'position' of the finite I in the manifold of the phenomenal experience (arising from the spontaneous activity of the mind). In this manifold, the finite I and nature represent the two complementary moments of the *absolute I* interacting with itself: "In I the ideal and the real grounds are one and the same, and this interaction between I and non-I is at the same time an interaction of the I with itself. It is able to posit itself as restricted by the non-I [and] it can posit itself as itself restricting the non-I." The expanding epistemic activity (of subjective positing) is no longer radically separated from the 'thing in itself' and objectivity of the natural world is reabsorbed into the activity of the I. The dynamic unity of the finite I, the non-I (nature), and the absolute I constitutes a "circle which finite I is able to extend into infinity, but can never escape."[40] In this new *Fichtean circle*, the finite I constantly 'touches' but never fully merges with the absolute I in its eternal "interaction … with itself." The logic of this process follows the dialectic of thesis-antithesis-synthesis that later finds its full development in the Hegelian dialectics. However, by contrast with the close-ended

---

[39] Fichte, G. J. (1982). *The Science of Knowledge* (P. Heath and J. Lachs. Eds. Trans.) (I, p. 225). Cambridge University Press.

[40] Fichte, G. J. (1982). *The Science of Knowledge*, I, pp. 280–281.

Hegelian dialectic, the open-ended Fichtean schema becomes the foundation of his open-ended philosophy of history.

Methodologically, as Cartesian rationalism finds support in analytical geometry and Leibniz's metaphysics—in infinitesimal calculus, Fichte's transcendentalism echoes the twentieth-century mathematical intuitionism and, more broadly, the existentialist problematic of human freedom, including the relationship between the transcendental and psychological ego. The act of freedom reaches the depth of unconsciousness, where the transcendental and empirical ego form the inseparable unity. In the state of intuition, "the mind lingers" between the two, setting the stage for unresolvable conflict between the absolute and finite I.[41]

Starting with the 1800th, the Fichtean transcendental methodology goes well beyond the demonstrative limits of 'clear and distinct' 'facts of consciousness' and develops into the system of objective idealism, where the Absolute is explicitly identified with God: God is all, and all is God. However, in opposition to the dogmatic (from the transcendental standpoint) pantheism of Spinoza and in a spirit of critical philosophy, for Fichte, God is identical with *knowledge* (rather than nature): "The existence of God is not a mere foundation, cause, or anything else of Knowledge … but it is absolutely Knowledge itself."[42] Considering that there is nothing but God, "a world has no existence but in Knowledge, and Knowledge itself is the World." The transcendency of God is reflected in knowledge as a "direct manifestation" of divine presence and "by means of Knowledge," it emerges as its "indirect manifestation" of the world.

---

[41] In the related line of thought, Brouwer finds the medium of the "synthesizing faculty" of (productive) intuition in time: "**Consciousness** in its deepest home seems to oscillate slowly, will-lessly, and reversibly between stillness and sensation… By a **move of time** a present sensation gives way to another present sensation in such a way that consciousness retains the former one as a past sensation, and moreover, through this distinction between present and past, recedes from both and from stillness, and becomes **mind**." Brouwer (2011)[1981]. "Consciousness, Philosophy and Mathematics" In *Brouwer's Cambridge Lectures on Intuitionism* (van Dallen. D. Ed.). Cambridge University Press. It would be interesting to compare Brouwer's distinction between consciousness and mind with a similar distinction between *chitta* and *manas* in Indian philosophy. We further address intuitionism in Chap. 5.

[42] Fichte elaborates on this ontology of knowledge in *The Facts of Consciousness* and summarizes his ideas in *The Theory of Science in its General Outline*. See Fichte, J. G. (1845–1846). *Sammtliche Werke* (J. G. Fichte (junior). Ed.) (Vol. II). Fichte's views can be compared with Maimonides' (1135–1204) contention that God is not distinct from his knowledge (but qualitatively distinct from human knowledge). See Maimonides. (n/d). *The Guide for Perplexed* (M. Friedlander. Trans.) (4th ed. Pt. III, ch. 20). E. P. Dutton & Co.

Fichtean understanding of the Absolute corresponds with St. John's "in the beginning was the Word (Logos), and the Word was from God, and the Word was God" and approaches the Neo-Platonic idea of the First Principle (*One-in-Itself*).[43] However, while God reveals itself as a *schema* or *knowledge*, in the Fichtean system, this relationship does not involve the (Neo-Platonic type of) emanation because God remains "the Absolute Being, transcending all Time."

In order to account for an internal dynamic (and historical unfolding) of knowledge, Fichte distinguishes between knowledge *in itself*, as a direct manifestation of God, and knowledge *for itself* that "becomes Self-Consciousness. As Self-Consciousness, knowledge acquires its own, peculiar, self-sustaining power, freedom, and activity," by which "Knowledge is constantly developing itself throughout Eternity [of the Absolute]" and "manifests itself as a *definite something*." The opposition of the *knowledge* and its outward manifestation makes it "unfold its own inward power… and in this progressive development we have the origin of Time," and of human history, we should add.[44]

By linking, somewhat inconsistently, the *transcendency* of the Absolute to the *transcendental* power of self-consciousness, Fichte tries to maintain the continuity of his earlier subject-oriented methodology focused on the historical dialectic of I and non-I. As a result of this ambiguity, his reasoning only recapitulates, albeit in a different form, the Kantian dualism of pure reason and the 'thing in itself'; paraphrasing Jacobi, we can enter Fichtean system only with the notion of the (transcendent) Absolute *in itself*, but we can remain within it only with the notion of the (transcendental) Absolute *for itself*.

Based on his dialectic of knowledge and being, Fichte articulated the principles of his philosophy of history in the 1806 work *Characteristics of the Present Age*, where he distinguishes between five historical epochs developing from the initial stage of 'prehistory' to the final realm of 'true history': "The Philosopher … [should] seek out the Idea of the Age… and must be able a priori to describe Time as a whole and all its possible Epochs."[45] The full realization of human freedom is an ideal goal of the last historical epoch. Due to the gaping chasm between the transcendency

---

[43] Compare Proclus (1992). *The Elements of Theology* (2nd ed., prop. 6). Clarendon Press.
[44] Fichte, I. G. (1977). *Characteristics of the Present Age* lecture 9, pp. 144–145.
[45] Fichte, I. G. (1977). *Characteristics of the Present Age* lecture 1, pp. 2–3.

of God and the world, this goal could only be accomplished as an infinite approximation toward the moral perfection of humanity.

Convinced that he overcame the dualism of his predecessors, Fichte, in his philosophy of history, does not seem as concerned as Leibniz and Kant with the 'harmony' between the ontologically distinct 'realms' of material necessity and ideal freedom. For Kant, nature (in itself) stands vis-a-vis human knowledge as a transcendent source of *theoretical* reconstruction, but for Fichte, it has a subordinate and an exclusively instrumental value, supporting the historical self-affirmation of humanity in the realm of necessity; indeed, "nature is only a border, only a negation, it should not have anything positive." On this account, the natural world is viewed as a passive object of human activity, something to be reabsorbed into the activity of the I—and humanity as its historical counterpart—and freely organized in accordance with the principles of reason.

At the same time, as a true son of the Enlightenment, Fichte underlines the role of science in human progress: "It is certainly true that …the satisfaction of human needs must become easier and easier, the efforts required for such a satisfaction must be decreasing, the fertility of land must be increasing, climate softening; <u>there must be made an infinite number of new discoveries and inventions</u> to increase means of subsistence and make life easier."[46] Fichte's open-ended conception of history and his emphasis on science as a major factor in historical progress followed the progressive spirit of the Enlightenment and became an inspiration for nineteenth-century post-Hegelian liberal thought.

In the last analysis, Fichte's ambitious attempt to overcome Kantian dualism and incorporate epistemology, philosophy of nature, and philosophy of history into the all-embracing *meta-science* of knowledge proved illusory. First and foremost, Fichte is incapable of integrating the effective causality of nature into his speculative dialectic of the absolute I. The disparity between freedom and nature is characteristic of both his earlier transcendental methodology and his later ontology of Knowledge. In this respect, Fichtean apriorism represents a step back from a more balanced dualism of Kant, making Fichte a target of critique for Schelling and Hegel, and the founder of German transcendentalism, Kant himself.

At the same time, ideas born in the lively and often bitter polemic between Kant, Fichte, Schelling, and Hegel opened the way to Schelling's

---

[46] Fichte, G. J. *Einige Vorlesungen Über die Bestimmung des Gelehrten* (1794) In *Johann Gottlieb Fichte's Sammtlihe Werke* (Bd. VI, s. 342).

conception of the Absolute as an *identity* of matter and spirit and its consequent development in the grand dialectical synthesis of Hegel. Moreover, Fichtean focus on human freedom and his historical optimism found new life in the social philosophy of the left-wing Hegelians and made a strong impact on Marx's philosophy of history and conception of philosophy as revolutionary praxis.

The Dialectic of History and the Myth of the State

Ironically, as Fichte claimed that he revealed the true spirit of transcendental methodology misinterpreted by Kant, Schelling insisted that "Fichte's subjective idealism… misunderstands itself." The key issue in the disagreement between Fichte and Schelling is the status of nature. Schelling's aspiration is to reconcile the formal causality of reason with the effective causality of nature. He departs from Fichtean subjectivism and insists that the genuinely monistic philosophy must begin with the *identity* of nature and spirit, explaining that for idealism as a system, "it is by no means adequate to claim that 'activity, life and freedom only are the truly real'; such statement would be compatible with Fichte's subjective idealism (which misunderstands itself)"; in addition, "it is required that the reverse also be shown, that everything real (nature, world of things) has activity, life and freedom as its ground or, in Fichte's expression, that not only is I-hood (*Ichheit*) all, but also the reverse, that all is I-hood."[47] At the same time, by contrast with the pantheism of Spinoza, "transcendental philosophy would be completed only if it could demonstrate this **identity** [of the ideal and real]… in **its principle** (namely the self)."[48] A faithful student of Kant, Schelling maintains that the totality (identity) of substance must be based on the "standpoint of reason."

The balancing act between subjectivism and realism makes Schelling a target of sharp critique for Fichte, his former mentor, who labels the principle of identity as nothing but a version of Spinozian dogmatism (i.e., uncritical materialism). At the same time, the principle of identity becomes a starting point for Schelling's friend and junior colleague, Hegel.

---

[47] Schelling, F. W. J. (2006). *Philosophical Investigations into the Essence of Human Freedom and the Related Matters* (J. Love and J. Schmidt. Trans.) (p. 22). State University of New York Press.
[48] Schelling, F. W. J. *The System of Transcendental Idealism* (p. 12).

Weighing on the 'dogmatism' vs. 'criticism' debate, Hegel initially sides with Schelling:

> It is necessary that Spinoza's substance is understood not as unmoved but as the intelligent, as a *form acting within itself with necessity*, so, that it is the creative principle of nature but at the same time knowledge and comprehension ... we need neither Spinoza's formal union nor the subjective totality of Fichte, but totality with the infinite form. We see this developing in Schelling's philosophy.[49]

Confronted with the long-standing problem of the identity between formality and actuality, Hegel rejects Cartesian intuition of 'clear and distinct' ideas and sees in *logic* the source of the "infinite form" revealing the totality of substance. From the standpoint of pure reason, it must be demonstrated how the unity between the formal causality (of reason) and its material base can be accomplished through *logical becoming*. Hegel explains:

> Thus, these two processes [nature approaching the ego and vice versa] are quite evident: realization of nature in order to reach the subject and realization of the I in order to reach the object. But *real* realization could take place *only in a logical way*, because it contains in itself pure thoughts; but the logical point of view is exactly what Schelling never arrived at in the presentation of his philosophy.[50]

Where Hegel parts ways with Schelling is in his thesis that the "logical way" (of activity) is identical to a "real realization" of nature. Hegel declares that a formal vehicle of the logical becoming and "real realization" of the Absolute is the *notion* that represents a synthesis of the rational *form* and its objective (material) *content*: "Only in its notion something obtains actuality," so that the *notion* is the "thought which is also the object (*Sache*) [of thought] as it is in itself."[51] The identity of the logical form and material actuality is succinctly expressed by Hegel's often-quoted remark that everything real (eventually) reveals itself as rational, and

---

[49] Hegel, G. W. F. (1963). *Lectures on the History of Philosophy* (E. S. Haldane and F. H. Simson. Trans.) (Vol. 3, p. 516, trans. modified). Humanities Press.

[50] Hegel *Lectures on the History of Philosophy* (Vol. 3, p. 518).

[51] Hegel, G. W. F. (2010). *The Science of Logic* (p. 29). (Trans. modified: 'Sache' is translated as 'object 'rather than 'fact'). Cambridge University Press.

everything rational is (to be actualized) as real. However, contrary to his own presuppositions, he 'dogmatically' posits the identity between formality and actuality, thought and matter as an empty 'notion' (lacking any actuality). Paradoxically, his 'dogmatic' panlogism recapitulates the "empty union" (of God's attributes) he attributes to Spinozism and signifies a breakup not only with Schelling but with the whole tradition of transcendental methodology.

Through logical reconstruction of the transcendency of the Absolute, Hegel does to Fichte what the latter did to Kant by allegedly penetrating the transcendent 'thing in itself' through the dialectic of productive imagination. By elevating Logic to the status of *ontology*, Hegel breaks the wall between the Absolute (especially as developed by Fichte in the 1800th) and its representation as knowledge and brings the Fichtean system to its logical conclusion: If *knowledge* is a "direct" schema of the Absolute (as claimed by Fichte), then its dialectical and historical reconstruction is a reconstruction of the Absolute itself, and the dialectic of *knowledge* does not represent just the *image* of the Absolute but the Absolute *itself*, as a totality with "infinite form." In this way, Hegel further develops the Fichtean *science of knowledge* into the system of *esoteric theology*; paraphrasing the Scripture, in the beginning was knowledge, and knowledge was from the Absolute, and knowledge was the Absolute. We will see in the next chapter how the theological core of his philosophy becomes a Trojan horse that triggered a split in the post-Hegelian movement.

Hegel's philosophy of history is intrinsically linked to his metaphysics. The structured unity of matter (content) and spirit (form) is a task to be accomplished through the dialectical *becoming* of their dormant *in itself* identity to the transparent *for itself* (objective) mode of the Spirit and, finally, the Absolute Idea. As the Spirit, the Absolute emerges as a *concrete universal* that finds its historical embodiment in the State, defined as an "actuality of the ethical Idea." Individual freedoms are subordinate to and can be realized only through the State. The State is a "living, organized Spirit" rather than a product of the arbitrary consensus (democratic or not) of the individuals. In this context, Hegel maintains that *civil society* represents only a multiplicity of the individual (and arbitrary) wills, lacking their essential unity; as such, civil society is only an 'external state,' where social existence is separate from its moral essence (ethical Idea). For Hegel, it is the focus of (liberal) civil society on human singularity and personal autonomy that does not allow it to express human essential freedom adequately. From this standpoint, he criticizes the *social contract* theory and argues that it is only through the centralized State that human freedom can find its true realization.

Hegel maintains that the moral essence of particular nations is expressed by the national states, which comprise, in their joint development, *world history*. In his early period, he argued that every nation has its unique historical path and "the absolute ethical totality is nothing other than a *people* [i.e., a nation]."[52] Reinforcing the supreme value of national self-identity (vs. that of singular individuals), and in opposition to Kant's ideal of eternal peace, he goes as far as insisting that "a war preserves the ethical health of peoples."[53]

While arguing that "each nation must first have gone through its own phases of [world] culture [*Kultur*]," he adds that this historical process prepares a nation for the time when it "intervenes in the universal world continuum." Moreover, "a principle which elevates it to universal dominion does not arise until its own distinct principle is applied to the rest of the unstable world system [*Weltwessen*]." Not unexpectedly, among the "distinct" people chosen to "intervene" in world affairs are Germans referred to as a "people from whom this universal shape [*Gestalt*] of the world spirit was born"; such attitude certainly sounds familiar in the context of the twentieth-century politics and remains relevant for contemporary political debates over the multipolar world order. In line with his political conservatism, Hegel views constitutional monarchy as a final form of the state for his native Prussia and the Germanic people as a whole.[54]

Hegel's historical schema of the two-fold 'end' of history and ideology, as well as his overall political conservatism, represent a step back from the historical optimism of the Enlightenment, even though Hegelian ideas may have, indeed, indirectly influenced modern liberal politics; for example, a Neo-Hegelian Alexander Kojev played a significant role in shaping up the political environment in the EU. While Fukuyama, referring to Kojev, unconvincingly tries to present Hegel as a precursor of modern liberalism, Hegel's philosophy of history does not seem to provide (to say

---

[52] Hegel, G. W. F. (1999). "On the Scientific Ways of Treating Natural Law, on its Place in Practical Philosophy, and its Relation to the Positive Sciences of Right" (1802–1803) In G. W. F. Hegel (1999). *Political Writings* (L. Dickey and H. B. Nisbet. Eds.) (p. 140). Cambridge University Press. In the mature form, these ideas are elaborated upon by Hegel in his 1821 Philosophy of Right; especially, pars. 257–261. Hegel, G. W. F (1942). *Philosophy of Right* (T. M. Knox. Trans.). Clarendon Press.

[53] Hegel, G. W. F. (1999), p. 141.

[54] Hegel, G. W. F. (1999). "The German Constitution" (1798–1802), pp. 63–64.

the least) a suitable background for the ideology of liberal democracy at either the national or international level. His system is mostly antagonistic to the Enlightenment (including Fichtean) focus on human activity and individual freedoms. In Hegel's conception of history, humanity becomes a passive witness to the close-ended historical dialectic of the Absolute, culminating in the (Prussian) State and Hegel's own philosophy. In his system of objective idealism, history is impregnated with reason and guided by a *providential plan* that *necessarily* requires its completion, the view succinctly expressed by Marx in a motto: 'Reason has always existed but not always in a reasonable form.'[55] At the same time, a radical reevaluation of the Hegelian philosophy in the mid-nineteenth century provided a background for developing European liberal tradition, while Hegel's dialectic of formality and actuality became the backbone of Marx's historical dialectic of labor.

From Dialectic of Reason to Historical Praxis

After Hegel died in 1831, his system came under attack from his former friend and ally, Schelling, who replaced Hegel at the University of Berlin. Schelling rejected Hegelian panlogism primarily for reducing the material causality of nature to the formal dialectic of the notion.

Looking for the identity of the material and ideal in the neutral plenum of the human lifeworld, Schelling turns to the study of mythological and religious experience. His analysis of world mythologies in *The Philosophy of Mythology* is closely connected to his earlier *Philosophy of Art*, where he characterizes mythology as a "unity of the infinite and the finite," at the core of which "lies the seed of the absolute."[56] Mythology provides the material allowing to present a synthesis of the intangible and tangible in a symbolic form (of philosophy and art). In this context, Schelling compares Greek mythology, where the infinity of the universe is represented in the concrete form of nature, with Christianity, where the Absolute is represented through the abstraction of the moral law. The focus of the former is to express the infinite through the finite (nature); the focus of the latter

---

[55] Hegel, G. W. F. *Encyclopedia of Philosophical Sciences* (II, par. 548–549). For an early influential discussion of Hegel's political views, see Haym, R. (2011). *Hegel Und Seine Zeit*. Nabu Press (Originally published in 1857).

[56] Schelling, F. W. J. (1989). *The Philosophy of Art* (p. 61). University of Minnesota Press. Cassirer makes numerous and favorable references to Schelling in Cassirer, E. (2020). *The Philosophy of Symbolic Forms*. Routledge (Originally published in 1923–1932).

is the opposite—to express the finite (nature) through the infinite (moral universe).

Contrary to the Hegelian apriorism, the structure of myth is defined not through the a priori dialectical laws but in the historical genesis of (collective) mythological consciousness, with reason understood as a historical phenomenon and its function and structure revealed through the phenomenology of historical consciousness. From this perspective, history represents a neutral medium for the whole body of human knowledge, and the critique of 'pure reason' is to be substituted (or at least complemented) by the critique of *historical reason*. On the surface, Schelling's position is similar to Marx's view of history as 'the only science we know.' However, unlike Schelling, Marx does not consistently maintain this methodological principle by declaring that human "history itself is a real part of *natural* history," reducing political and ideological 'superstructure' to its material 'base' and thus opening a way to accusing him of the naturalistic fallacy.[57]

In examining mythological consciousness, Schelling goes beyond the speculative tenets of transcendental tradition, trying to reveal the historical genesis of the "mystery" of *productive imagination,* one of the central notions of Kant's architecture of reason and the Fichtean dialectic of the I. In mythological consciousness, defining human *lifeworld*, the subjective activity of productive imagination merges with its primordial archetype of the Absolute, where the sacred and profane, ideal and material modes of human existence form an inseparable unity. The phenomenology of the lifeworld, a genuine medium of the historical process, defines its internal dynamic, the scope of historical experience, and the extent of historical time. On this account, Schelling firmly opposes the purely logical constructions of "senseless progress without limits" as a form of fruitless historical apriorism. He refers to infinite progress as a "symbol of faith of our time" that turns historical time into an empty form for a prearranged association of historical events.[58] On the other hand, Schelling also rejects

---

[57] Marx, K. *Economic and Philosophical Manuscripts* In Marx & Engels (2010). *Collected Works* (V. 3, pp. 303–304). Lawrence and Wishart.

[58] The philosophical significance of mythology for the philosophy of history is underlined already by Fichte in the *Characteristics of the Present Age* (Lecture 9). Schelling's reflections on the philosophy of history are in Schelling, F. W. J. (2018). *Einleitung in die Philosophie der Mythologie* (Lecture 10). Forgotten Books. His understanding of mythology as a form of collective consciousness is in tune with the theological critique of David Strauss to be examined in the next chapter.

Hegel's close-ended dialectic of history as based on the speculative formalism of the dialectical method. In the spirit of this critique, any attempt to create an all-embracing ideological or historical conceptual framework ('liberal' or otherwise) that claims to find the 'end' of either ideology or history could be viewed as a form of shallow and dangerous historical apriorism that is doomed to fail (in a more or less violent way).

While the debate over Hegel's system, in general, and his philosophy of history, in particular, was initially focused on the theological core of his philosophy, Schelling's theological exegesis did not significantly impact post-Hegelian philosophical development. In the spirit of the age, Marx, associated with the left-wing critics of Hegel, declares that "mythology... disappears with a real mastery over the forces of nature."[59] Most importantly, the disagreement about the relationship between Hegel's system and Christianity eventually outgrew into the complete revision of his dialectical method and philosophy of history. Rejecting the historical implications of Hegel's system, the Hegelian left (Young Hegelians) turned to the Fichtean philosophy of freedom as an inspiration for their liberal ideology and political activism. Their radical critique turned Hegelian dialectic into the instrument of social critique and historical change. It is not the self-realization of the Absolute but the force of philosophical and social critique that becomes an engine of progress, inevitably leading to the expansion of social freedoms in civil society.

Marx's critique of Hegel eventually goes well beyond the half-hearted liberalism of Young Hegelians. In line with Schelling's analysis of the limits of theoretical reason, Marx proclaims that the task of philosophy is not just to explain but to change the world. Accordingly, he rejects subjective interpretations of the Hegelian method by Bruno Bauer and his circle and replaces the power of philosophical critique with the revolutionary praxis based on the historical dialectic of labor. Confronted with the problem of reconciling subjective human activity with the natural human environment, Marx construes (collective) labor as a genuine synthesis of the universal and material modes of human activity. As we show in the consequent two chapters, the inconsistency between the formal (universal) and material aspects of Marx's 'inverted' Hegelian dialectic of labor leads to the inconsistency between his close-ended dialectic of technology and the

---

[59] Marx, K. & Engels, F. (1975–1995). *Collected Works* (Vol. 46 (1), p. 47). International Publishers. From a strictly historical standpoint, Marx's historical naturalism could be itself viewed as a product of the mythology of the Enlightenment.

open-ended conception of history. Historically, the new *Marxian historical circle* recapitulates the problematic intrinsic to the Cartesian circle of formality and actuality and to the circular Hegelian dialectic of "substance acting within itself."

In the next chapter, we examine the evolution of the left-Hegelian movement, from the theological critique of Hegel by David Strauss to Bruno Bauer's 'Fichteanization' of Hegelian dialectic in the conception of philosophy as social critique, Feuerbach's inversion of Hegelian 'esoteric theology' into the project of social anthropology and Moses Hess' theory of *socialism*. In this context, we discuss a strong, though diverse, influence of the group's key members on Marx's conception of *historical materialism*.

CHAPTER 3

From Social Anthropology to the Historical Dialectic of Labor

Man is the measure of all things: existing as they exist
and non-existing as they do not exist
—Protagoras

In the professed humanism of the Enlightenment tradition, the system of politics is inseparable from the system of ethics guided by rational principles; in this spirit, Marx characterizes Kant's revolution in philosophy as the "German theory of French revolution." In a similar vein, Fichte, in the *Contribution to the Correction of the Public's Judgment on the French Revolution* (1793), argues for the right of the people to end their 'social contract' with the state. Underlining a connection between progressive politics and the revolutionary spirit of his philosophy, he writes in the 1795 letter to Jens Baggesen: "My system is the first system of freedom. Just as France freed man from external shackles, so my system frees him from the fetters of things in themselves… Indeed, it was while I was writing about the French Revolution that I was awarded by the first hints and intimations of this system."[1] Joining his senior colleagues, Hegel enthusiastically supported the ideals of freedom and human rights advanced by the French Revolution.

---

[1] Fichte, J. G. *Briefwechsel* (Bd. 2, pp. 449–450).) (hrsg. Von Hans Schulz).

However, the early enthusiasm faded due to the ensuing atrocities of the Jacobin terror, the Thermidorian Reaction, and the ultimate failure of the revolution.[2] As a result, the progressive philosophical spirit of Fichte and Hegel gave way to a more measured conception of political freedoms, with a new appreciation of the Kantian distinction between the arbitrary will, based on natural inclinations, and the universal (general) will, based on the principles of reason. At the height of this movement, Hegel's system, enormously popular at the time, culminates in the conservative political theory of the State as a secular embodiment of 'ethical Idea' and 'divine will,' advancing the principle of the essential unity between philosophy and religion, the State and the Church.

After Hegel's death, the central issue that came to the forefront of the controversy over his heritage was the ambiguous relationship between his system and traditional Christianity. The right-wing Hegelians believed that the master's system was in full conformity with the traditional Christian dogma. The position of the Hegelian center was to have the Bible reinterpreted and fully integrated into the Hegelian system. Finally, the left-wing argued that the biblical narrative is full of inconsistencies and that traditional religions may have only practical value, as was already suggested by Spinoza. The debate over theological subtleties eventually triggered the split among Hegel's followers on the whole corpus of Hegelian philosophy. The historically influential group of left-wing Young Hegelians, theologians and philosophers by education and social critics by vocation, was the first to attack the theological core of Hegel's philosophy, gradually leading them to a complete revision of Hegel's system and its historical and political implications.

The crucial role in this movement belongs to David Strauss, initially a member of the Hegelian center, who undertakes a thorough critical examination of the Gospels. He targets the conspicuous contradictions in the narrative of Jesus as a God-like personality and tries to reconcile the Hegelian system with religious faith. The consequent critique of Ludwig Feuerbach goes beyond the more measured theological exegesis of Strauss and proclaims that "the secret of **theology** is **anthropology**, but **theology** itself is the secret of **speculative philosophy**."[3] In this way, he replaces

---

[2] Drawing a parallel between the French and Russian Revolutions, Trotsky labels Stalin's rise to power as *Thermidor*.

[3] Feuerbach, L. (1972). "Preliminary Theses on the Reform of Philosophy" In *The Fiery Brook: Selected Writings of Ludwig Feuerbach*. (p. 103). Anchor Books.

Hegelian universality of the Absolute with the universality of man and turns Hegelian philosophy into the atheistic project of *social anthropology*.

The anthropological approach of Feuerbach opens the way to the social and political critique of Bruno Bauer, Arnold Ruge, and others, who turn to Fichte as their inspiration and emphasize the primacy of reason as a driving force of historical progress. The political radicalization of the Hegelian left was accompanied by the reevaluation of what they saw as inconsistency between the close-ended character of Hegel's system, including his conception of history, vs. his open-ended dialectical method. Overall, the historical significance of the Young Hegelians lies not so much in the depth of their theological or epistemological analyses but in their role in transforming philosophy into the instrument of social critique and action. For them, in Marx's words, "the criticism of religion [turns] into the *criticism of law*, and the *criticism of theology* into the *criticism of politics*."[4]

Due to the lasting influence and multifaceted character of his worldview, Marx occupies a unique place in the post-Hegelian movement. His conception of history is influenced by Bauer's social activism, Feuerbach's project of social anthropology, and Hess's views on socialism. In a close association with Hess and Engels, Marx outlines a program of socialism as a comprehensive theory of a humane society that goes far beyond the halfhearted liberalism of Bauer or Ruge. On the philosophical front, his objective is to develop a new system of *humanism*, surpassing both materialism and idealism. At its core is a synthesis of Feuerbachian anthropological naturalism with the *inverted* Hegelian dialectic. The result is a circular dialectic of objective human activity (labor) and objective human essence. The new *Marxian circle* is crucial for understanding how the quest for objectivity shaped Marx's anthropocentric ontology and prophetic historicism.

As we explore the views of the Young Hegelians, our objective is not to present a detailed assessment of the post-Hegelian movement but to explain its influence on the evolution of Marx's thought. His association with and the consequent critique of the Hegelian left played a crucial role in the development of historical materialism, plagued by the inconsistency

---

[4] Marx, K. (1975). *A Contribution to the Critique of Hegel's Philosophy of Right* (1844) In Marx, K. *Early Writings.* (p. 243) (R. Livingstone and G. Benton. Trans.). Vintage Books.

between the open-ended conception of history and the circular dialectic of labor.[5]

FROM DIVINE PROVIDENCE TO HUMANITY

The advances in the *rational* reconstruction of theology were gradually undermining its very foundation, making religious faith increasingly susceptible to the critique of reason. Already, Spinoza's *Theologico-Political Treatise* presents a devastating analysis of the biblical narrative, attacking the key principles of the Judeo-Christian tradition. For the *pantheism* of Spinoza, God, as the only substance, represents a complete unity of essence and existence. Comparing Cartesian dualism to the pantheism of Spinoza, Feuerbach explains that for Descartes, the two (material and thinking) substances depend on God only in their *existence* but not in *essence*, while for Spinoza, "the dependence of existence—which is really, taken **in itself**, not a dependence—becomes also the dependence of essence."[6] As a totality of essence and existence, God "poured all his creative power into a system of forms, and he is bound to these forms. That is what pantheism means."[7] From this perspective, God is not a personality *transcending* the world (with all the ensuing problems of creation and emanation) but is a power immanent to the material world.

For theology, pantheism becomes a Trojan horse: If God and Nature are one—that is the principle of pantheism— then God, "bound" by the lawful "system of forms," is a legitimate subject matter for scientific investigation. In other words, religious narratives of the divine historical 'becoming' (of the essence to its existential realization) may present, at best, an incomplete or distorted image of reality, prone to critical scrutiny in the same way that empirical facts are verified and analyzed by natural science. In this spirit, Spinoza argues that a detailed analysis of the Bible reveals so many inconsistencies that what remains is a moral core common to all major religious traditions: love your neighbor as yourself. Then, each form of organized religion has only practical significance as a code of conduct developed by and for a particular nation.

---

[5] An extensive discussion of the Young Hegelians can be found in McLellan, D. (1969). *The Young Hegelians and Karl Marx*. Frederick A. Praeger. See also Hook, S. (1958). *From Hegel to Marx*. The Humanities Press. A brief account of the movement is in Kolakowski, L. (1978). *Main Currents of Marxism* (Vol. 1). Oxford University Press.

[6] Feuerbach, L. *The History of Modern Philosophy* (Chap. 8, prop. 82).

[7] Tillich, P. (1951). *Systematic Theology* (pp. 235–236). University of Chicago Press.

Moreover, the very idea that religion could and should be tested on the scale of reason was perceived by many as a challenge not only to theological circles but also to the political status quo. Unsurprisingly, in 1656, Spinoza was condemned by the Jewish religious establishment and excommunicated. His *Theologico-Political Treatise*, published in 1670, was banned by the government four years later despite the relatively tolerant intellectual environment of the Netherlands (his homeland).

Spinoza's radical critique of the Bible was unacceptable not only to the Synagogue (or the Church, for that matter) but even to his intellectual comrades and sympathizers. Even Leibniz, despite his close contact with and respect for Spinoza, distances himself from the Philosopher; and Kant is careful to distinguish between traditional and rational theology, keeping the rational reconstruction of Christianity encapsulated within the boundaries of 'pure reason'; the spirit of a measured approach to religion is vividly expressed by the title of his seminal work, *Religion within the Bounds of Reason Alone*. In spite of a conciliatory attitude towards organized religion, Kant was reprimanded by the Prussian government after the second edition of the aforementioned book had been published. Fichte, Kant's celebrated follower, began his career with an essentially atheistic work, *Attempt at a Critique of All Revelation*, in which he openly defied the religious establishment and defended intellectual freedom. Though this work made his name known to the public and helped to launch his academic career, he was later forced to leave the academic position in Jena and moved to Berlin amidst the accusations of atheism.

As Kant's followers' original support of the French Revolution gave way to a more reserved political outlook, their early critique of religion gave way to attempts to unite philosophy and religion. Fichtean philosophy developed into a form of *esoteric theology*, with the Absolute conceived as a creative power of God defining the historical *telos* of humankind. In the system of Hegel, the speculative and theological components merge into a virtually inseparable unity, securing his philosophy a firm support from the state and almost an official status of secularized religion.

Hegel argues that "logic is to be understood as a System of Pure Reason or a Realm of Pure Thought. **This realm is the Truth as it is, without husk in and for itself...** The content [of logic] **shows forth God as he is in his eternal essence before the creation of Nature and of a Finite**

Spirit."[8] Behind this highly speculative guise is an ultimate ambition to penetrate the inner core of the Absolute (God), something that neither Kant nor Fichte deemed possible. The abstract ontology of the Absolute becomes a shell for the theistic core of Hegel's system in general and his philosophy of history in particular. In this framework, human history represents an 'autobiography of God,' a revelation of his absolute freedom; and the social fabric of human life becomes a mirror reflecting God's eternal essence at the consequent stages of its existential self-realization. By integrating God into the historical dialectic of the Absolute, Hegel completely integrates theology into philosophy, specifying, in the *Lectures on the Philosophy of Religion*, that "philosophy is itself, in fact, worship... identical with religion."[9]

The very unity of philosophy and theology implies that the historical reconstruction and critique of biblical narrative entail a conceptual critique of Hegel's system itself. Conversely, the critical analysis of Hegel's speculative philosophy amounts to deconstructing its theological core. On this background, the theological treatise of David Strauss (1808–1874), *The Life of Jesus* (1835), proved to be a pivotal point in the radical reassessment of Hegel's philosophy. Strauss' analysis of the biblical narrative sharply divided Hegelianism, with theology becoming a conceptual battleground for Hegel's followers, from left to right. Says Strauss:

> To the question whether and how far the Gospel history is contained as history in the idea of the unity of divine and human nature, there are three possible answers: namely that from this concept either the whole Gospel narrative or only a part of it or finally that neither the whole of it nor a part of it can be deduced as history from the idea. If these three answers or directions were each represented by a branch of the Hegelian School, then one could, following the traditional simile, call the first direction the right, as one standing nearest to the old system, the third left and the second the center[10]

---

[8] Hegel, G. W. F. *Science of Logic* (p. 60) (Johnston W. H. and Struthers L. G. Trans.). George Allen & Unwin Ltd.

[9] Hegel, G. W. F. (1962). *Lectures on the Philosophy of Religion* (Vol. 1, p.20). Routledge and Kegan Paul.

[10] Strauss, D. (1841). *Streitschriften* (Bd. 3, s. 95). Tubingen; quoted in McLellan. (1969). p. 3.

What is at stake is a relationship between the actuality of the historical process and "the idea of the unity of divine and human nature," exemplified by a singular individual—Jesus. The problem is the intrinsic ambiguity of the relationship between the speculative and theological elements of Hegel's system. Developing ideas of the late Fichte, Hegel states that "God is Spirit, the *activity of pure knowledge*" that "eternally produces himself in the form of his Son." Thus, the Son is a realization of God's creative power, with the two related by the eternal logos (knowledge). In turn, this speculative (esoteric) schema has its counterpart in historical Christianity, where "the relation between [God] Father and [God] Son is expressed in the form of organic life, and is used in popular or figurative sense… and, accordingly, never entirely corresponds to the truth that is sought to be expressed."[11] Hegel, in fact, retains (under the speculative guise) the Kantian disparity between the 'religion within reason only' and the traditional Christianity.

As Fichte undermined his philosophy of nature by eliminating Kantian thing in itself, Hegel undermined his philosophy of religion by the *symbolic* linking of Christian trinity to the logical 'moments' of the dialectical unfolding of the Absolute. It is the tension between the speculative idea of Son vs. its "popular" historical instantiation as an actual individual that becomes a starting point for the theological critique of Strauss. In traditional Christianity, Jesus is an *existing individual,* and the biblical narrative is an actual account of human history. This position is unacceptable for Strauss because he sees the Gospels as full of inconsistencies that are difficult (at best) to reconcile. By contrast, if the figure of Jesus is interpreted only as an essential *symbol* of 'the unity of divine and human nature' (i.e., in a "figurative" sense as Hegel implies), then "neither the whole [of the Gospels] nor a part of it can be deduced as history from the idea [of such a unity]."

Following this route and trying to maintain a centrist balance between philosophical rigor and religious faith, Strauss turns to Kant's theological exegesis: "According to Kant… it ought not to be made a condition of salvation to believe, that there was once a man who by his holiness and merit gave satisfaction for himself and for all others"; but "it is humanity… in its entire moral perfection that could alone make a world the object of divine Providence, and the end of creation." In other words, within the Kantian rational reconstruction of theology, Jesus, as an empirical

[11] Hegel, G. W. F. (1968). *Lectures on the Philosophy of Religion* (Vol. 3, p. 11).

individual, cannot represent an ultimate synthesis of the divine essence and the 'outward experience'; such belief would be outside of the 'limits of reason.' However, God may serve as a regulative idea of the moral (practical) reason for *humanity* in its *infinitely* unfolding historical progress towards the moral kingdom of ends, a position also consistent with a Fichtean philosophy of history.

At the same time, Kant's attempt to reconcile 'rational' and 'popular' theology and accommodate Christian dogma is undermined by the intrinsic dualism of his system. With a rigid separation between reality (in itself) and the phenomenal world (for us), human life is torn between the contingency of its historical existence and the ideal realm of moral ends. Elaborating on the core problem of Kant's rational theology, Strauss says: "This idea of humanity... proceeds from his [God's] essence, and is therefore no created thing, but his eternal Son" but, continues Strauss, "this ideal remains essentially confined to the reason, because it cannot be adequately represented by any example of outward experience." Conversely, for humanity confined to the world of phenomena, there is no room for divine experience, and "if the idea has no corresponding reality, it is an empty obligation."[12] The proper solution should be based on the "idea which has an existence in reality, not in the mind only, like that of Kant."[13]

The *infinity* of the assumed historical conversion of the causality of the natural world to the teleology of moral ends only underlines the gaping chasm between the divine essence and its 'outward realization,' highlighting the historical dilemma Strauss (and Kant) confronted. By the same token, Kant's reference to the hidden 'secret plan of nature' as a source of the elusive harmony between social life and its assumed moral telos is also unacceptable for Strauss due to the underlying dualism of nature and human freedom.

On this background, Strauss aims to discover an adequate medium capable of connecting divine essence to its existential realization in history. His solution is to shift the historical unity of man and God from one individual (Jesus) to the medium of "the myth-making consciousness of the Christian community." As a species concept, the myth assumingly allows Strauss to overcome Kantian dualism and salvage Christianity from the inconsistencies of the *New Testament*. He explains in the *Life of Jesus*:

---

[12] Strauss, D. F. (1860). *The Life of Jesus Critically Examined* (Vol. 2, p. 888) (Marian Evans. Trans.). Calvin Blanchard.

[13] Strauss. (1860). Vol. 2, p. 895.

It is not the way in which the Idea realizes itself to pour out its whole fullness in one example... rather it likes to spread out its reaches in the multiplicity of examples which mutually complete themselves... When thought of as belonging to an individual, a God-man, the qualities and function that the teaching of the Church attributes to Christ are contradictory, but in the species they live in harmony. Humanity is the unity of both natures, finite spirit remembering its infinity[14]

For Strauss, the human *species*, as a bearer of divine nature, is not a mere collection of empirical individuals (even conceived as a succession of generations) but a 'myth-making consciousness,' revealing itself in the course of human history. From this perspective, the Gospels express the collective consciousness and, as such, should be a subject matter of theological (and hermeneutical) exegesis that combines symbolic interpretations with historical data rather than being a target for empirically based historical analysis and critique (as held by Spinoza).

While Kant insists (consistently maintaining disparity between the noumena and phenomena) on the principal difference between traditional and rational theology and, in this way, sidelines the significance of biblical narrative, Strauss tries to preserve the integrity of theological discourse by focusing on 'pure' phenomenology of myth-making consciousness. With mythology as the *essence* of Christianity, the Gospel history does not need to be rationally deduced and all factual inconsistencies reconciled. In this way, Strauss separates theology from the speculative elements of the Hegelian system and effectively rejects an idea of the essential identity between philosophy and theology. Then, the Bible, in general, and the account of Jesus in particular are not directly linked to the idea of the Absolute and its dialectical self-realization in history.

Strauss' interest in mythology was consistent with the intellectual milieu of the epoch; in this respect, we should mention the pioneering works of Friedrich Schlegel (1772–1829) and Schelling on mythology and religion. After Hegel's death in 1831, Schelling replaced him in Berlin and delivered a series of lectures in the philosophy of mythology. As was discussed in the previous chapter, Schelling's objective was to trace the divide between spirit and matter through the historical genesis of mythological

---

[14] Strauss, D. F. *Das Leben Jesus* (Bd. 2, pp. 743ff); quoted in McLellan (1969). p. 92. Marion Evans uses the term 'race' instead of the 'species': Strauss (1860), p. 895.

consciousness.[15] Strauss' approach follows Schelling's interpretation of myth as an *identity* (synthesis) of culture and nature, ideal and the real, essence and existence. In the twentieth century, Strauss' treatment of mythological consciousness echoes Thomas Mann's historical reinterpretation of the biblical narrative in his masterpiece, *Joseph and His Brothers.* Mann conceives biblical characters not as concrete individuals but as instantiations of mythological archetypes. On this account, the symbolic figures of biblical forefathers are as part of modern history as modern history is symbolic. For Mann, the dormant humanistic core of this symbolism remains alive amidst one of the darkest periods in European history, and in mythological archetypes of the biblical narrative he sees an essential core of humanity, the most secure historical foundation of directional progress.[16]

Historically, Strauss' centrist position proved unstable, opening a critical Pandora's Box for the opportunities to attack and ultimately abandon both the biblical narrative and its Hegelian esoteric counterpart. Eventually, the attempts to reconcile the Hegelian philosophy with traditional religion lead to the outright rejection of the whole biblical corpus. If "the secret of myth is to be found in the way the community lives… [and] the history of the religious myth is the history of the community that nurtured it," then 'phenomenology of mythological consciousness' also could and should be studied in its concrete historical forms, including the Gospel narratives, which must be examined in the light of historical facts—that was the criticism leveled against Strauss by Bruno Bauer, a trained theologian and a onetime teacher of Marx.[17] On this background, the developing post-Hegelian liberalism hailed the disintegration of all forms of mythological and religious experience, opening a way to the culture of technological totalization, with its desacralization of human relation to nature and deterioration of ecological values.[18]

---

[15] Schlegel, K. W. F. (1808). *Über die Sprache und Weisheit der Indier.* Hook, S. (1958). *From Hegel to Marx,* p. 82.

[16] Mann, T. (2005). *Joseph and His Brothers* (pp. 43–119). Alfred A. Knopf. Consider, especially 'Prelude' (Decent into Hell) and 'The Stories of Jacob.'

[17] Hook, S. (1958). *From Hegel to Marx,* p. 83.

[18] The lesson Marx assumingly learned towards the end of his career. See, for example, the discussion of Marx's 1860th notebooks in Saito, K. (2023). *Marx in the Anthropocene.* Cambridge University Press. The thoughtful discussion of technological vs. traditional 'totalization' can be found in Don Ihde. (1990). *Technology and the Lifeworld.* Indiana University Press.

The Dialectic of Reason as Social Critique

Rejecting Strauss' (and Schelling's) appeal to the phenomenology of collective consciousness, Bruno Bauer (1809–1882) insisted that the explanations of biblical inconsistencies through the concept of mythology were entirely arbitrary because myth itself is a historical epiphenomenon and the so-called sacred texts are human artifacts which must be studied as such. As a theologian, Bauer concludes that of the four Gospels, only one is original, while the other three are compilations. As for the original Gospel, it was composed by a particular individual and represented a personal and essentially distorted account of history rather than a product of the collective consciousness of humanity as a species.

Bauer's theological critique is accompanied by the critique of Hegel's speculative philosophy. Bauer claimed that his 'Fichteanized' Hegelian dialectic revealed the genuine meaning of Hegel's philosophy, as Fichte claimed that he revealed the genuine meaning of Kant's philosophy by disposing of the thing-in-itself. His central charge is an alleged contradiction between the close-ended structure of Hegel's system and its open-ended dialectical method, with a key implication that Hegelian dialectic calls for an open-ended, infinite unfolding of human history. Aligning with Bauer, Engels, Marx's closest associate, points to the contradiction between the Hegelian system and the allegedly open-ended character of human history:

> In accordance with traditional requirements, a system of philosophy must conclude with some sort of absolute truth. Therefore, however much Hegel, especially in his Logic, emphasizes that this eternal truth is nothing but the logical, or, the historical process itself, he nevertheless finds himself compelled to supply this process with an end, just because he has to bring his system to a termination [at] some point or other[19]

Let us explore Bauer's position in more detail. In the words of McLellan, Bauer "radically parted from Hegel... in his conception of the dialectic" and "changed the dialectic into a purely negative one: instead of the later state of affairs being the expression of former one-sided ones in their completion, it now negated them and was, in extreme cases, their exact

---

[19] Engels, F. (1990). *Ludwig Feuerbach and the End of Classical German Philosophy* (1886) In Marx, K. & Engels, F. *Collected Works* (V. 26, p. 360). International Publishers.

opposite."[20] Accordingly, neither divine nor human essence is being revealed 'for itself' in either the human pursuit of truth or human history. Elaborating on Bauer's conception of dialectic, McLellan further observes: "Whereas Hegel had tried very hard to preserve a unity between thought and being, Bauer, and the Young Hegelians in general, gave the precedence to thought, that of the human subject. There only remained the subjective side of Hegel's philosophy of identity."[21] As a result, for them, the subject of history is human self-consciousness (instead of the Hegelian Absolute), and philosophy becomes a form of *social and historical praxis*: **"theory is praxis**, and for that very reason most dangerous, far-reaching and destructive. It is the revolution itself." In line with the ideals of the Enlightenment, the power of reason is proclaimed a driving force of historical progress.

Bauer construes historical praxis in the predominantly ideal, theoretical mode and insists that "a theoretical principle must not merely play a supportive role, but must come to the act, to practical opposition, to turn itself directly into *praxis* and action... And so philosophy must be active in politics, and whenever the established order contradicts the self-consciousness of philosophy, it must be directly attacked and shaken."[22] Such criticism, guided by the normative system of moral principles, is the 'most dangerous' and efficient instrument of historical change. The open-ended dialectic of philosophical criticism creates a potentially infinite historical dynamic; paraphrasing Bernstein, in Bauer's conception of history, the road to the moral kingdom of ends is everything, and the final goal is nothing.

Despite the progressive spirit of Bauer's critique, his claim that Hegel's system is inconsistent with his dialectical method seems unfounded. Bauer de facto substituted the circular Hegelian dialectic with an open-ended dialectic of Fichte, claiming, at the same time, that he preserved the true (revolutionary) spirit of Hegel's doctrine (as we remember, Fichte made a similar claim regarding Kant and Schelling—regarding Fichte). In his assessment of Bauer, Kolakowski observes, "the notion that it is possible to salvage from the Hegelian dialectic the idea of eternal progress while

[20] McLellan (1969). p. 52.
[21] McLellan. (1969). pp. 52–53.
[22] Bauer, B. (1983). *The Trumpet of the Last Judgment over Hegel the Atheist and Antichrist* (1841): *An Ultimatum* (Die Posaune des jungsten Gerichts über Hegel den Atheisten und Antichristen: ein Ultimatum) In Stepelevich, L. S. (Ed.). *The Young Hegelians: An Anthology* (pp. 183–184). Cambridge University Press.

jettisoning the conservative idea of an ultimate goal is analogous to a philosophy which, confronted with the contradiction between God's omnipotence and man's free will, should abolish God and claim that it preserved the essence of Christianity...The whole burden of Hegel's criticism... of the 'bad infinity' or notion of unending growth was that any phase of development can only be understood in relation to a final state, without which so-called progress is merely eternal repetition."[23] In other words, the close-ended character of the historical process is inseparable from the Hegelian dialectical method; and Bauer, with his open-ended dialectic of subjective spirit, simply abandons the Hegelian philosophy and turns 'back to Fichte.'

In spite of all its flaws, the idea of permanent criticism as an instrument of historical change contributed to the rapid radicalization of the political atmosphere in Germany. In particular, Arnold Ruge (1802–1880) was especially active in opening ideas of the Hegelian left for public discussion. In the 1840th, he founded a political party and edited a newspaper, advocating liberal reforms.[24] In line with Bauer's critique, he writes in 1840 that "the unity of historical and purely spiritual process... leaves nothing but the world of reason" and maintains that the most important political principle of the time is "autonomy of the spirit, namely in science—development of rationalism and in state politics—liberalism."

Ruge struggled for freedom of speech, viewing political pluralism as both a prerequisite and an outcome of social freedoms. "The oracle of our time," he proclaims, "is a revolution of the European humanity."[25] His writings made a noticeable impact on the Hegelian left and beyond. Ruge was engaged in political activities across Europe, including participating in the European Democratic Committee. His liberalism was more concrete than the program of pure criticism of Bauer's circle and gravitated to the increasingly radical politics of Moses Hess and Karl Marx, with whom, despite their disagreements, Ruge collaborated in the 1844 *German–French Annals* (The *Deutsch–Französische Jahrbücher*) they co-edited.

---

[23] Kolakowski, L. (1978). *Main Currents of Marxism.* Clarendon Press (Vol. 1, pp. 402–403).

[24] On Ruge see Hook, S. (1958), Brazill, W. J. (1970), Stepelevich, L. S. (Ed.). (1983). Bauer's criticism of Hegelian essentialism could be compared to the critique of essentialism by Karl Popper. Popper, K. R. (1965). "Prediction and Prophecy in Social Sciences" In Popper, K. R. *Conjectures and Refutations: The Growth of Scientific Knowledge* (pp. 336–347). Harper & Row.

[25] *Gallische Jahrbücher.* 1841, 1, 2.

Marx, associated with the Hegelian left in his early period, initially aligned himself with Ruge's political liberalism and supported Bauer's philosophy of self-consciousness. In his 1843 response to Ruge, he writes, "Philosophy has now become secularized and... there can still be no doubt about the task confronting us at present: the *ruthless criticism of the existing order*."[26] In his *Doctoral Dissertation* (1841), Marx reiterates Bauer's position that philosophy has to "recognize human self-consciousness as the highest godhead" and concurs with Bauer that "the praxis of philosophy is itself theoretical." In the spirit of German idealism, he declares that philosophy "is the sort of critique that measures individual existing things by their essence and particular realities by the Idea."[27] This early attitude is well expressed by his often-cited statement that "reason has always existed, but not always in a reasonable form."

The influence of Bauer on Marx continues even after Marx's breakup with the Young Hegelians. The criticism of Bauer in *The Holy Family* (1845) and *The German Ideology* (1845) is directed only at a particular aspect of Bauer's views, while the progressive spirit of Bauer's critique remains with Marx for the rest of his career. Siding with Bauer's atheism, he proclaims that "criticism of religion is the premise of all criticism," emphasizing that all proofs of the existence of God prove only the existence of human self-consciousness. Marx emphasizes that the criticism of religion implies the criticism of the alienated forms of self-consciousness as they arise in the political, economic, or other realms of human life; religion, in this sense, is the most fundamental form of human self-alienation.

While eventually moving beyond the philosophy of pure criticism, Marx and Engels (his friend and a close associate by that time) recognize the importance of its political implications: "Just as destructive criticism in Germany, before it had progressed in Feuerbach to the consideration of real man, tried to resolve everything definite and existing by the principle of self-consciousness, destructive criticism in France tried to do the same by the principle of equality."[28] The programmatic statement in *The German Ideology* (written by Marx in collaboration with Engels and, possibly, Hess) that "we know only one science—the science of history" also demonstrates

---

[26] *Deutsch-Französische Jahrbücher*, 1844; https://www.marxists.org/archive/marx/works/1843/letters/43_09-alt.htm

[27] See McLellan. (1969). p. 71.

[28] Marx, K. Engels, F. (1990). *The Holy Family*, p. 30.

the influence of Fichteanized 'negative dialectic' of history advanced by the Hegelian left.[29]

By 1844, the split between Bauer and Marx became conspicuous and involved the two key issues. First, Marx sharpens Bauer's position by proclaiming that through critical activity, "the world is becoming philosophical" and philosophy, as a political force, is "becoming worldly"; consequently, he concludes, "with philosophy's becoming worldly, the realization of philosophy coincides with its disappearance."[30] In this way, he decidedly moves beyond a purely theoretical attitude to the world, which is characteristic of both Fichte and Bauer, and begins to develop his own approach to historical praxis. In this context, Marx increasingly recognizes the importance of political and economic struggle based on the material and social conditions of human life and, more specifically, socioeconomic class relations. As it becomes clear from the *Economic and Philosophical Manuscripts* (1844), *The Holy Family* (1844), and especially *The German Ideology* (1845), his goal is to find an *objective* mode of material praxis commensurate with the objectivity (aka *universality*) of human social and natural existence. In this respect, *The German Ideology* signifies Marx's complete breakup with an exclusively theoretical attitude to the world. In this work, he forcefully puts forward the idea that social consciousness is based on mundane conditions of human life: "In direct contrast to German philosophy which descends from heaven to earth, here we ascend from earth to heaven... Life is not determined by consciousness, but consciousness by life."[31]

Second, the sharp split among the Hegelian left was mainly due to the growing influence of Feuerbach's anthropological inversion of Hegelian theological essentialism.[32] As McLellan observes, "by the end of 1843 the influence of Feuerbach was supreme among all those left-wing Hegelians who did not attach themselves to Bruno Bauer: Marx, Engels, Hess, Ruge."[33] Bauer and his circle were highly critical of the concept of *species*

---

[29] See McLellan. (1969). p. 50, f.1. McLellan emphasizes the connection of Bauer's and Marx's views and criticizes Sidney Hook who allegedly "fails to take into account the earlier period of Bauer's writings [and] confuses it with the period of 'pure criticism.'"

[30] Karl Marx. (1978). *Selected Writings* (pp. 13–14) (D. McLellan. Ed.). Oxford University Press.

[31] Karl Marx (1978), p. 164.

[32] For a comprehensive study of Feuerbach see Wartofsky, M. W. (1977). *Feuerbach*. Cambridge University Press.

[33] McLellan. (1969), p. 104.

and how Feuerbach, Marx, and some other group members used it.[34] In particular, Bauer claimed that Feuerbach de facto replaced Hegelian esoteric theology with a new religion of abstract human essence isolated from any political or social context. The advantage of his own critique of Hegel, argues Bauer, is that it uproots Hegelian essentialism by targeting not only religious alienation but all alienated forms of human self-consciousness.

This overall well-founded criticism was also directed at Marx. Marx's essentialism of that period is apparent in his understanding of man as a *species being*. In the 1844 *Paris Manuscripts*, he uses the notion of species in the attempts to overcome the one-sided character of idealism and materialism and develop a theory of *humane society*, with the primary objective to highlight the contradictions between the universality of human *species life* (as an expression of human essence) and alienated forms of human *existence* in the capitalist society. Moreover, it would be a mistake to say that only the 'early' Marx used the concept of *species*, while the 'mature' Marx gave it up. The conceptual apparatus of 'mature' Marx is inseparable from the essentialism of his early works, including his signature concepts of communism, proletariat, and alienation. Without the notion of human essence, Marx's theory of communism loses its humanistic appeal of the upcoming 'realm of freedom,' where all forms of social alienation are to be superseded, and it is Feuerbach's social anthropology that gave an impetus to Marx's quest to find a new, `universal` mode of human activity commensurate with man as a universal (species) being.

From Social Critique to Social Anthropology

Strauss' theological exegesis paved the way not only for Bauer's social critique but also for the anthropological inversion of the Hegelian system by Ludwig Feuerbach (1804–1872). While Strauss targeted the concept of trinity, Feuerbach disposed of God as a transcendent entity altogether: "The task of the modern era was the realization and humanization of God—the transformation and <u>dissolution of theology into anthropology</u>."[35] As an ideological expression of the 'modern era,' "the new philosophy makes **man, together with nature** as the basis of man, the **exclusive, universal, and highest object** of philosophy; it makes **anthropology,**

---

[34] See, for example Bauer, B. (1983). "The Genus and the Crowd" (1844) In Stepelevich. (1983), pp. 198–207.

[35] Bauer (1983), par. 1

together with physiology, the universal science."[36] What constitutes man as a subject of universal science is a power of self-consciousness that reveals him *for himself* as a *species* (universal) being. Feuerbach calls this mode of self-consciousness "consciousness in the strictest sense":

> What is the essential difference between man and the brute? The most simple, general, and also the most popular answer to this question is—consciousness—but consciousness in the strict sense; for the consciousness implied in the feeling of self as an individual... cannot be denied to the brutes. <u>Consciousness in the strictest sense is present only in a being to whom his species, his essential nature, is an object of thought.</u>[37]

Feuerbach emphasizes an intrinsic connection between human self-consciousness, human essential (species) nature, and an *intersubjective* mode of human *species existence*: "Man is himself at once I and thou; he can put himself in the place of another, for this reason, that to him his species, his essential nature, and not merely his individuality, is an object of thought."[38] So, it is my social relatedness to another individual, as it emerges in the act of reproducing the 'I-Thou' relation in my consciousness, that constitutes me as a *species-being*.[39] By contrast, an animal cannot enter into universal (species) relations with either nature or other animals because it is conscious of itself only as an (isolated) individual and does not possess consciousness 'in the strictest sense.'[40]

As a matter of fact, Feuerbach's argument represents a socialized form of Cartesian *cogito*: *I think* of myself as a universal (social) being, therefore *I exist* as a universal (social) being, i.e., my *species* (truly human) existence is constituted by the act of thought; this means that as *universal* human beings, men can relate to each other only *in thought*. The abstract character of this argument reveals the weakest point of Feuerbach's

---

[36] Bauer (1983), par. 54.
[37] Feuerbach, L. (1881). *The Essence of Christianity* (p. 1) (Marian Evans. Trans.). London: Trubner and Co.
[38] Feuerbach (1881), p. 2.
[39] Feuerbach, L. (1881), pp. 1–2. For Feuerbach, 'man is a universal being' means 'man is a species-being.' Feuerbach, in his use of term species (*Gattung*), follows Strauss: "Is it at all possible that a species realizes itself in **one** individual...?" etc. See Feuerbach, L. (1972). "Towards a Critique of Hegelian Philosophy" (1839) (p. 56) In *The Fiery Brook Selected Writings of Ludwig Feuerbach* (Hanfi, Z. Trans.). Anchor Books.
[40] Feuerbach, L. (1881), p. 1.

anthropological project: a chasm between the ideal character of *human essence* and the natural (including socioeconomic) mode of *human existence*.

In defining the species nature of man through his identity with another, Feuerbach follows Hegel, for whom "the relation of the Kind [species] is the identity of the individual self-feeling in an entity which at the same time is another independent individual." In this universal "relation of the Kind," an individual recognizes his "identity with another individual" and, consequently, recognizes himself as *essentially human*. However, Feuerbach departs from Hegel, for whom "an individual is species **in itself**, it is not species **for itself**."[41] In Hegel's system, only God (Absolute) is a *species* both *in* and *for itself*. In other words, God represents a *totality* where *existence* is identical with its *essence*; by contrast, man is a species in itself but not a species for itself; this means that human existence and essence are distinct, and man only potentially, through his unity with God, may achieve an identity of essence and existence. Paradoxically, at this point, Hegelian *essentialism* reaches the core of the existentialist attitude to the world: human existence precedes essence; to emphasize this point, Hegel remarks, in one of his early works, that human decision-making must always be concrete (allowing a concrete existential choice to reveal its rational core).

By contrast with Hegel, Feuerbach conceptualizes man as 'Man-God,' a whole, *total being* or as a being given *for himself* in the totality of his social and natural existence: "**The real in its reality and totality, the object of the new philosophy**, is the object also of a **real** and **total** being. The new philosophy therefore regards as its **epistemological principle**, as its **subject, not the ego, not the absolute**—i.e., abstract spirit, **in short, not reason for itself alone**—but the **real** and the **whole being of man**."[42] In this way, theology (and, by extension, Hegel's speculative philosophy) is transformed into *anthropology*.

Trying to extend the universality of human social self-consciousness to the totality of human life, Feuerbach presents a generalized form of his social *cogito* in order to to accommodate the mechanism of mutual self-recognition to human relation towards nature: "Man is nothing without

---

[41] Hegel, G. W. F. (1929). *Science of Logic* (Vol. 2, pp. 413–414) (W. H. Johnston, L. G. Struthers. Trans.). The Macmillan Co.

[42] Feuerbach, L. (1972). "The Principles of the Philosophy of the Future." (1843) (par. 50) In *The Fiery Brook: Selected Writings of Ludwig Feuerbach*.

an object... consciousness of the objective is the self-consciousness of man. We know the man by the object, by his conception of what is external to himself... this object is his manifested nature, his true objective *ego*."[43] That is why, in the framework of social cogito, to recognize himself as a universal social being, man needs, vis-à-vis himself, another man as a mirror of his own image. It is by consequently internalizing their identity that man is able to identify his *universal human nature*.

By further applying the reciprocal character of social (man to man) relations to the natural world, Feuerbach implies that relatedness to *natural* objects, "external" to man, would constitute him not only as a *universal social* being but also as a *universal natural* being. However, in the framework of Feuerbach's anthropology, man cannot enter into the reciprocal activity of self-recognition with natural objects in the same way he does so with other men (albeit also in an abstract way). As a result, man remains, at best, an allegedly *universal social being* but not a *universal natural being* and, consequently, not a *total being*.

Apart from the problem of mutual self-recognition, what Feuerbach defines as 'natural' human relations is confined to the contingent domain of love, friendship, or sympathy, which does not represent humanity in its essential (species) nature; on this account, in his natural life, man does not seem to differ from 'brute,' and human society is restricted to nothing more than an abstract association of self-conscious 'monads' (a la Leibniz), each representing a full image of the social whole. Here, the inconsistency between the material mode of their existence and the ideal mode of their 'participation' (in the Platonic sense) in the harmony of the (social) universe remains unresolved.

Feuerbach's anthropological turn is often considered an essential step towards Marx's materialist conception of history. As observed by Alfred Schmidt: "Feuerbach turned Thought, Spirit, from an absolute Subject into a human quality alongside other natural qualities." In this context, Schmidt underlines the contrast between Kant's transcendentalism and Feuerbach's materialism, stating that "in Feuerbach, who stood at the end of this whole movement of thought, man became the unique theme, precisely as an empirical and natural being." However, it would be a mistake to overestimate Feuerbach's split with Kant because as transcendental philosophy maintains a gap between "the transcendental Ego and the empirical-psychological Ego" so does Feuerbach fail to link the ideal mode

---

[43] Feuerbach, L. (1881), pp. 3–4.

of human essence to the empirical mode of human existence and, thus, overcome the old Kantian dichotomy of man as a contingent object in the *realm of necessity* vs. man as a universal subject in the *realm of freedom*.[44]

In the last analysis, the failure to understand man as both a universal social *and* natural being undermines Feuerbach's conception of man as a *whole* being and a subject of the new universal science. Nevertheless, his attempt to define human essence in social terms and link man and nature through the dialectic of self-objectification paved the way for Marx's dialectic of labor as technology-mediated self-externalization.

Social Anthropology as Socialism

Along with Bauer's critical philosophy and Feuerbach's social anthropology, an important place in the development of Marx's signature theory of *historical materialism* belongs to German socialist Moses Hess (1812–1875) and Polish writer August Cieszkowski (1814–1894), who sought to "abolish the difference between action and philosophy."[45] In this way, "philosophy was to become an act of will instead of merely reflection and interpretation, and was to turn itself towards the future instead of the past."[46] While the Polish writer was associated with the Hegelian left, it is hardly possible to attribute to him any direct influence on Marx. However, he could influence Marx through Hess, who saw in Cieszkowski's philosophy of praxis a step from the abstract Bauer's criticism to *socialism*.

The relationship between Hess and Marx is a complicated matter. Hess held Marx in great esteem and, most likely, was under the spell of Marx's ideas. In one of his letters (September 2, 1841), he says, "Dr. Marx... combines the deepest philosophical seriousness with the most biting wit. Imagine Rousseau, Voltaire, Holbach, Lessing, Heine and Hegel fused

---

[44] Schmidt, A. (1973). *The Concept of Nature in Marx* (p. 25). London: NBL.
[45] On Cieszkowski see Liebich, A. (1979). *Between Ideology and Utopia: The Politics and Philosophy of August Cieszkowski*. Dordrecht: D. Reidel. On Hess: Berlin, I. (1959). *The Life and Opinions of Moses Hess*. For a brief account of their influence on Marx see Kolakowski, L. (1978). *Main Currents of Marxism* (Vol. 1, pp. 85–88, 108–114) (P. S. Falla. Trans.). Claredon Press. Another representative ontology: Fried A. and Sanders, R. (Eds.). (1993). *Socialist Thought: A Documentary History*. Columbia University Press
[46] Cieszkowski A. von. (1838). *Prolegomena zur Historiosophia* (p. 129). Berlin. Quoted also in: Kolakowski. (1978). p. 85.

into one person... and you have Dr. Marx."⁴⁷ However, by that time, Marx had written nothing substantial, and the letter conveyed hardly more than Hess' general impression from their first meeting.

Hess joined the Hegelian left in 1838–1839 but soon moved beyond the 'anarchy of liberalism' and abstract philosophical critique, arguing for the importance of social emancipation through revolutionary struggle. He emphasizes the importance of economic relations as a material foundation of society and proclaims that atheism and communism are the future of a genuinely humane society.⁴⁸ Like Marx and other Young Hegelians, he finds inspiration in Feuerbach's critique of religion but construes the Feuerbachian conception of anthropology as a theory of *human society* or *socialism*:

> Theology is **anthropology**. That is the truth [discovered by Feuerbach], but it is not the **whole** truth. The nature of man, it must be added, is social, involving the cooperative activity of all individuals for the same ends and interests. The true theory of man, the true humanism is the theory of **human society**. In other words, **anthropology is socialism**.⁴⁹

While for Feuerbach, the idea of God represents human essence in its alienated form, Hess, in one of his most influential works, 'On the Essence of Money,' links *alienation* to socioeconomic reality and the power of capital: "What God is for the theoretical life, money is for the practical life of the inverted world: the alienated power of men, their reified activity";⁵⁰ to the extent that individuals relate to each other through money, they are being perceived as objectified products of hostile market forces instead of free actors evaluated by their *essentially human* (species) nature.⁵¹ Needless to say, this approach is in tune with Marx's views on social

---

⁴⁷ Quoted in McLellan (1969). p. 145. This letter is often quoted by the orthodox Marxists who deny any influence of Hess on Marx; see Oizerman (1962). *Formirovanie Filosofii Marksizma* [The Origins of Marxist Philosophy] (p. 102). Moskva: Izdatel'stvo Sotzial'no Ekonomicheskaya Literatura.

⁴⁸ Hess, M. (1993). *The Philosophy of the Act. In Socialist Thought: A Documentary History.*

⁴⁹ Quoted in Hook, S. (1958). *From Hegel*, p. 196.

⁵⁰ Hess, M. (1961). *Philosophische und Sozialistische Schriften. 1837–1850*. Berlin. Quoted in McLellan *The Young Hegelians*, p. 157.

⁵¹ The next chapter explores a difference between objectification as a form of natural self-expression vs. alienation rooted in unfair economic mode of production and distribution. For the related discussion see Marcuse, H. (1973). "The Foundations of Historical Materialism" In *Studies in Critical Philosophy* (J. de Bres. Trans.). Beacon. Marcuse, H. (1954). *Reason and Revolution*. The Humanities Press.

alienation and, in particular, with his conception of the fetishism of commodities.

Hess influenced the founders of 'scientific communism' in various ways. As McLellan observes, "the similarities between Hess' article 'On the Essence of Money' (1843-1844) and Marx's 'Judenfrage' are remarkable and can only be accounted for on the supposition that Marx copied heavily from Hess' essay presuming it would not be published."[52] Hess also authored a part of *The German Ideology*, and, in 1847, he published a short work, *Consequences of the Revolution of the Proletariat*, close in content to the famous *Communist Manifesto* (1848). In this work, as Sidney Hook puts it, one can find "the theory of concentration and centralization of capital, the theory of increasing misery, the theory of overproduction... the doctrine that the collapse of capitalism is inevitable, and the view that the development of revolutionary consciousness is a simple and direct outgrowth of economic distress."[53] He maintains that the material foundation of social progress is the development of 'big industry' and the "surplus of unutilized instruments of production that it is extremely easy to produce through them abundantly all that we require."[54] This idea is echoed by the founders of Marxism in the early 1840th and reiterated later by Marx in The *Grundrisse* and the *Capital*. Hess' influence is especially conspicuous in Marx's understanding of humanism in 1844 *Economic and Philosophical Manuscripts*, where he links the advancement of humane society to the emancipation of humanity from the *sense of having*. Moreover, Hess allegedly played a direct role in Engels' conversion into a communist, an assessment that the orthodox Marxists contest.

Marx approaches Hess' conception of alienation in the influential article 'On the Jewish Question' (1843), a review of Bauer's work with the same title and elaborates on the idea of *true humanism* in the context of a Hegelian distinction between the nominalist conception of *civil society* and an essentialist conception of *social humanity*. "The standpoint of the old materialism is civil society; the standpoint of the new is human society or

---

[52] McLellan. (1969), p. 155.
[53] Hook, S. (1958). *From Hegel to Marx*, p. 204.
[54] Hess, M. (2004). *Consequences of the Revolution of the Proletaria*t. In Hess, M. *The Holy History of Mankind and Other Writings* (Shlomo Avineri. Ed.) (pp. 128–136). Cambridge University Press.

social humanity."⁵⁵ In civil society, individuals, unable to realize their genuine human essence, are being entangled in the contingent net of material necessity; they relate to each other in the singularity of their alienated social and natural existence, unable to freely express themselves as universal (species) social and natural beings. Following Hess, Marx points out that the social estrangement of man from his "inner wealth" in the capitalist society is based on the *sense of having*, and to be freed from the sense of having, "human being had to be reduced to this absolute poverty in order that he might yield his inner wealth to the outer world (On the category of 'having,' see Hess in the Einundzwanzig Bogen [Hess' *The Philosophy of the Act*])." This task can be accomplished through "the abolition of private property [which] is therefore the complete emancipation of all human senses and qualities, but it is this emancipation precisely because these senses and attributes have become, subjectively and objectively, human."⁵⁶

Overall, Marx's humanistic aspirations, including the early program of "emancipation all human senses" and the later emphasis on unlimited technological expansion as a basis for the transition to socialism, could be attributed (among other factors) to his collaboration with Hess. The conception of social humanity, developed by Marx in collaboration with Hess and Engels, represents the core of his historical universalism. For an accurate assessment of Marx's historical materialism, it is essential that the fundamentals of his 'early' humanism are intrinsically linked to the 'mature' socioeconomic analysis of the *Capital*. Indeed, the emancipation from the 'sense of having' and its socioeconomic expression in private property is a path to a truly *humane society* devoid of alienation and exploitation of one man by another. To the extent that capitalism inevitably leads to the pauperization of the masses and deepens the tension between the rich and poor, it digs its own grave by creating a universal historical force allegedly devoid of the sense of having—the *proletariat*—and opening a way to the upcoming 'realm of freedom,' where "the conflict of man's individual, material existence with his species-existence" is to be finally resolved.⁵⁷

Moreover, the distinction between social humanity and civil society provides a theoretical foundation for Marx's critique of Feuerbach's

---

[55] Marx "Theses on Feuerbach," theses 6 & 10; in Karl Marx. (1978). *Selected Writings*. p. 156 (McLellan. Ed.).

[56] Marx. (2010). *Economic and Philosophical Manuscripts of 1844*. In Marx & Engels. *Collected Works*. (V. 3, p. 300). Lawrence & Wishart.

[57] Marx. "On the Jewish Question" In Karl Marx *Selected Writings* (p. 62) (McLellan. Ed.).

abstract understanding of human nature. While for Feuerbach, human essence remains an "abstraction inherent in each single individual," Marx eliminates the disparity between the proclaimed universality of man and the historical dynamic of human life by introducing an operational notion of human essence as an *ensemble of social relations*. In this way, he breaks the wall—ontological, epistemological, and historical—between the universality of man and natural conditions of human existence, between the humane *realm of freedom* and the material *realm of necessity*. The universality of man is now linked to and defined through the activity of productive labor and the corresponding cluster of socioeconomic relations within the "legal and political superstructure... to which correspond definite forms of social consciousness."[58] In this context, Marx Wartofsky observes that individuals are "no more concrete than the relations in which they stand to each other; a point Marx was to drive home."[59]

While such a move seemingly links human universality to social life, it relativizes the understanding of man as a universal being by reducing this universality to historically transient socioeconomic conditions. This methodological reductionism can be illustrated by comparing Marx's 'inversion' of Feuerbach with the nominalist critique of Feuerbach by Max Stirner (1806–1856), who is sometimes viewed as a transitional figure in the development of Marx's historical materialism. For example, a noted Austro-Marxist Max Adler (1873–1937) and more recently McLellan claim that Marx's critique of Feuerbach was influenced by Stirner. In the article written on the occasion of Stirner's hundred-year birthday, Adler firmly places Stirner "between Feuerbach and Marx" and credits him with dethroning any form of ideological fetishism (including religious one). Adler argues that by reducing (social) ideology to the interests of concrete individuals, Stirner opens the way to a materialist understanding of history.

However, Stirner's critique of Hegel and Feuerbach goes much further than that of Marx, though, in many respects, in a different direction. In his magnum opus, *The Ego and His Own* (1844) (also translated as *The Unique and Its Property*), Stirner attacks the abstract character Feuerbach's anthropological *realism* (in the scholastic sense) that heavily relies on the concept of man as a universal (species) being and consistently maintains a strict *nominalist* position, devoid of any abstractions. He views Feuerbach's

---

[58] Marx. K. (1977). *A Contribution to the Critique of Political Economy* (1859) (Preface) Moscow: Progress Publishers.
[59] Wartofsky, M. W. (1982). *Feuerbach* (p. 426). Cambridge University Press.

anthropology as an expression of religious consciousness, with an abstraction of God replaced by that of man. In this context, he rejects the notion of essence (*species*) as an ideal archetype innate to the human or divine mind and revealed 'for itself' in its dialectical or historical becoming; nor does he agree with either the (Hegelian) understanding of essence as a 'relation of the kind,' or a corresponding (Feuerbachian) I-Though 'species relation' of socialized individuals. Stirner argues that Feuerbach's anthropology, allegedly focused on man as a whole being, de facto retains abstract Hegelian essentialism while ignoring the singularity of each particular individual living within his 'own' natural circle of interests. For Stirner, the notion of essence refers to nothing more than a retrospective, static 'snapshot' of human existence; and, by taking such a snapshot (something that inevitably occurs in individual and historical consciousness), this existence is being negated, superseded, and reduced to a pure abstraction; expressed in the lingo of twentieth-century existentialism, this position could succinctly expressed by the motto 'existence precedes essence.'

As the concept of the *human essence* is nothing but an empty abstraction, so is the idea of *humanity*. In the introduction to his famous book, Stirner asks a rhetorical question that flies in the face of Marx's historical essentialism: "Is not humanity's cause a purely egoistic affair?"[60] From his strictly nominalist standpoint, individuals are *more concrete* than any 'ensemble of their social relations.' In this way, Stirner tries to underpin the intrinsic contradiction between the claim of the supreme value of an individual as a universal (species) creature and the political appeal to sacrifice the individual *here-and-now* existence for the sake of a forever retreating future. As a result, the singularity of human existence is dissolved in a quasi-humanistic rhetoric rooted in the speculations of transcendental philosophy. His uncompromising attack on essentialism, either in its theoretical or practical (moral/social) mode, and the unwavering focus on human singularity make him, along with Kierkegaard, an early precursor of Nietzsche and the twentieth-century existentialist philosophy. From this perspective, Stirner's philosophy of man signifies a decisive breakup with the Enlightenment tradition in general and German classical idealism in particular.[61]

---

[60] Stirner, M. (2017). *The Unique and Its Property* (p. 26). Underground Amusements Pub.
[61] In this context, Engels could substitute Feuerbach by Stirner in the title of his famous book *Ludwig Feuerbach and the End of German Classical Philosophy*.

Although Marx may have been influenced by Stirner's critique of social abstractions, the central problem for him is not the notion of species per se but the abstract way in which human species life is construed by Feuerbach. His central charge is that, for Feuerbach, human species relations are separated from the socioeconomic reality, and, as a result, man is isolated from any historical context.[62] It is to remediate this problem that Marx gives an 'objective' form to the concept of human essence by reducing it to an ensemble of social relations. In this way, the universality of man as a species being is transferred to the proletariat as a *universal class*. However, the balancing act between methodological nominalism (a la Stirner) and realism (a la Feuerbach) proves unsustainable, with far-reaching implications for the conception of 'scientific communism.' The chasm between the ideal of man as a universal being and the proletariat as his socioeconomic alter ego undermines the viability of the transition from the ideal of *humane* society to its practical implementation. This problem found a vivid historical expression in the Russian October 1917 Revolution, the consequent Thermidor of Stalinism, and the final dissolution of the Soviet system. To paraphrase Jacobi again, we cannot enter Marx's humanism without the notion of a universal being, but we cannot stay within the Marxist political practice with this notion.[63]

Emancipation of Human Senses and Ontology of Labor

For the founders of Marxism, the central task in the reassessment of Hegel and his left-wing followers is to identify a universal (objective) mode of human social and natural activity and, thus, understand man in the *totality* of his spiritual, social, and natural existence. Addressing this problem in the 'Theses on Feuerbach,' Marx points to the conceptual *incompleteness* of both the preceding materialist and idealist traditions: "The chief defect of all hitherto existing materialism… is that the thing, reality… is conceived only in the form of the **object of contemplation**, but not… subjectively"; furthermore, "the **active** side was developed abstractly by idealism [and]

---

[62] "Feuerbach… is compelled to abstract from the historical process… and to presuppose an abstract—isolated—human individual." Marx. "Theses on Feuerbach," thesis 6.

[63] Lucacs, in his last, unfinished work, emphasizes an importance of the notion of species for Marx's thought: Lucacs, G. (1986). *Prolegomena Zur Ontologie des Gesellschaftlichen Seins*. Luchterhand.

Feuerbach... does not conceive human activity itself as **objective** activity."[64] Feuerbach becomes a target of this critique to the extent that his anthropology combines both of the weaknesses: on the one hand, the (social) universality of man is construed through the ideal activity of self-consciousness, and on the other, man as a natural being is conceived, in a 'contemplative' way, through the contingent emotions of 'love and friendship.' A formal juxtaposition of idealism and materialism makes Feuerbach's attempt to combine the universal and material nature of man *inconsistent*. Marx and Engels elaborate on this issue in their historical materialism manifesto, *The German Ideology*:

> Feuerbach... conceives of men not in their given social connection, not under their existing conditions of life, which have made them what they are, he never arrives at the really existing active men, but stops at the abstraction 'man', and gets no further than recognizing 'the true individual, corporeal man' emotionally, i.e. he knows no other 'human relationships' of 'man to man' than love and friendship, and even then idealized[65]

The inconsistency of Feuerbach's anthropocentric inversion of Hegel becomes a starting point for the development of Marx's philosophy of *humanism*, allegedly surpassing the one-sided view of the world by materialism and idealism. Marx finds the universal (objective) way in which people relate to each other *and* nature in *productive labor*. As both a *universal* and material mode of human praxis and the basis on which the rest of human social life emerges as a *superstructure*, labor represents man as a *universal social* being, a *universal natural* being, and, hence, a *total being*.

While the emphasis on labor as a material form of human activity is understandable and consistent with the general direction of Marx's thought, the essentialist claim of universality betrays the Hegelian roots of his methodological stance. As we will see in the next chapter, Marx de facto substitutes an ideal activity of mutual self-recognition with the self-referential character of labor by claiming that the dialectic of *self-reference and objectification* allows man to reveal himself as a being *for himself* and, as such, an objective (universal) being. In this way, Marx recapitulates the Feuerbachian dialectic of social self-consciousness, itself virtually identical to the Hegelian dialectic of the 'kind'; as a matter of fact, it is the

---

[64] Marx, "Theses on Feuerbach," thesis 1.
[65] Karl Marx. (1978). *Selected Writings*, p. 175 (McLellan. Ed.).

Feuerbachian *social cogito* that lies at the foundation of Marx's conceptions of labor as an *objective* human activity and of man as a *universal* being. Following these presuppositions, he construes the historical dialectic of technology-based labor through the speculative pattern of *self-externalization*.

Most importantly, Marx does not demonstrate *a posteriori* how the material character of labor is related to the dialectic of self-reference and objectification that constitutes man as a universal being. He adopts a self-referential pattern of labor *a priori* in order to present labor as an objective (aka universal) human activity. Following this logic, Marx construes technology as a tool mediating human intercourse with nature: instead of seeing his image in another individual, man recognizes himself as an objective being in tools, instruments, and machines, which allegedly represent an objective (externalized) expression of his 'essential powers.' In this conceptual framework, technology becomes a vehicle of human material self-externalization and an objective 'mirror' allowing man to perceive himself as a universal natural being. This approach to the mediating function of technology closely follows one presented in Hegel's *Science of Logic*.

In her analysis of the inconsistency of Marx's early 'humanist' view of human nature with his later 'scientific' conception of labor, Amy Wendling argues that "labor becomes mere labor power, a point where energy is transferred, but little more." Consequently, "human nature... is fully assimilated to the natural world, whose activity is indifferent from—and interchangeable with—that of machines, animals, and nature itself."[66] In fact, the implied inconsistency between Marx's early view on human nature and his conception of labor is a reflection of a deeper problem: the dual character of labor as it was formulated in the 1844 *Paris Manuscripts* and developed in his later writings, including the *Capital*.

As we argue in the next chapter, the inconsistency of Marx's *humanism* in general and his understanding of human nature, in particular, is rooted in the inconsistency of his conception of labor as both the universal and material human activity. In his dialectic of technology-based labor—and, by extension, the materialist conception of history—Marx tries to combine two seemingly conflicting ideas of man as a *material* being immersed in his natural environment and a *universal* being destined to become a 'master and possessor' of the universe. As was shown in the previous

---

[66] See Wendling, A. E. (2009). *Karl Marx on Technology and Alienation* (p. 3). Palgrave Macmillan.

chapter, this problem goes back to the Cartesian dualism of man as a thinking substance and a material being, which, in turn, can be traced to the scholastic dichotomy of formality and actuality.

The next chapter examines how Marx develops his dialectic of labor, utilizing the 'inverted' Hegelian dialectic, where an anthropocentric interplay of labor and objective human essence replaces the ideal self-realization of the Absolute. This analysis helps shed light on the conceptual foundations of his humanism and understand whether an open-ended conception of history—and the postulate of infinite technological progress as its key constitutive element—is consistent with a circular dialectical pattern of technology-based labor.

CHAPTER 4

The Dialectic of Labor and the Limits of Technological Growth

Ontology Recapitulates Technology

Marx presents his comprehensive worldview as a liberating philosophy of "consistent naturalism or humanism… distinguished from both idealism and materialism [which], at the same time, constitutes their unifying truth."[1] He explains in the 1844 *Paris Manuscripts* that a "fully developed naturalism equals humanism… it is the genuine resolution of the conflict between man and nature and between man and man—the true resolution of the strife between existence and essence, between objectification and self-confirmation, between freedom and necessity, between the individual and the species." At the core of this humanistic manifesto is the elimination of the *sense of having* and the transformation of all human senses, leading to the creation of the new man of the future.

The clue to the practical implementation of this worldview lies in changing human attitude to private property, which "has made us so stupid and one-sided that an object is only ours when we have it." In this context, Marx's objective is to demonstrate how the historical development of labor relations leads to the abolition of private property. He maintains that the *essence* of private property ("the essence of wealth") is the

---

[1] Marx, K. (1978). "On the Jewish Question" In Karl Marx *Selected Writings* (p. 181) (McLellan. Ed.). Oxford University Press. Marx, K. "Theses on Feuerbach" thesis 1.

idea of labor and, in its modern form, of *industry*. With the growing role of capital as a universal form of property and a "world-historical power," technology-based labor (industry) becomes a universal driver of historical change.

Marx claims that the classical political economy does not see the internal contradiction between *subjective* and *objective* aspects of property, i.e., between *labor* and *capital*. He argues that the resolution of the dynamic relationship between capital and labor can occur only through overcoming the distorted attitude to labor in a capitalist society, turning it into a means of liberation rather than the alienation and enslavement of man. In its progressive historical role, labor is represented by the *proletariat*, a socioeconomic group reduced to absolute poverty and, thus, becoming a leading force in the transition to communism—the historical stage of "emancipation of all human senses and qualities."[2] It is through the emancipated labor that man can enter into an equal relationship with fellow humans and the coextensive relationship with nature and, in this way, realize his essence as a truly *human, total* being. The dialectical interplay of labor, technology, and nature remains a recurrent theme for Marx, from his 'early' works, especially *The Economic and Philosophical Manuscripts* (1844) and *The German Ideology* (1845), to the 'mature' *Grundrisse* (1858) and his opus magnus, *Capital*.

The principal means of ameliorating the human condition and the historical transition to communism the founders of Marxism see in the unbounded scientific and technological appropriation of the world. The central role of technology in the conception of *historical materialism* is underlined by the bold contention that the preceding history of mankind had been the 'history of industry.' While Marx recognizes that expanding material production leads to the corresponding expansion of human 'wants,' he views an infinite cycle of production and consumption as a vital element of the transition to and sustainable development of a communist paradise, where everyone could have it according to his needs. From this perspective, any constraints on technological progress would constitute a significant problem for the productivist, in essence, theory of scientific communism.[3]

---

[2] The citations above are from 1844 *Paris Manuscripts*. Marx & Engels. (2010). *Collected Works* (Vol. 3, pp. 291–300).

[3] For 'realm of freedom' and 'realm of necessity' see: Marx, K. (1984). *Capital* (Vol. 3, p. 820) (F. Engels. Ed.). International Publishers.

Methodologically, Marx's dialectic of labor and technology recapitulates, in its essential elements, the structure of the Hegelian, closed-ended dialectic. In his dialectical schema, technological progress is conceived as a process of continuous human *self-externalization*, allowing man to realize himself as a *being for himself* and, as such, a *universal* (species) being. However, the circular dialectic of self-externalization proves inconsistent with the postulate of infinite technological progress and, by extension, the open-ended conception of history. Historically, this inconsistency reflects the two competing, Hegelian and Fichtean, elements in Marx's historical synthesis. While the dialectic of labor is based on the anthropocentric inversion of the Hegelian dialectic, the conception of open-ended progress is influenced by the Fichteanized idea of history, advanced by Marx's former allies, the Young Hegelians. The split in the Marxist movement appears as an inevitable outcome of the multifaceted and not always consistent structure of Marx's philosophy, combining, to a different degree, the elements of Kantian apriorism, Fichtean philosophy of freedom and Hegelian dialectical essentialism.

In the first part of the chapter, we explore Marx's anthropocentric ontology, including his understanding of human universality, and consider whether the self-referential dialectic of labor is consistent with, central to scientific communism, the ideologeme of infinite technological growth. In the second part, we address a controversy between the 'orthodox' and 'revisionist' Marxism and conclude with a discussion of the early twentieth-century Russian philosopher, scientist and political activist A. Bogdanov (Malinovsky). His highly original interpretation of Marx influenced by the brewing scientific revolution and popular at the time Machian *empiriocriticism* turns Marx's anthropocentric philosophy into the comprehensive project of universal organizational science. The examination of his allegedly scientific proof of 'inexhaustible' human creativity leads us to the problematic of the contemporary 'limits to growth' and 'end of science' debates explored in the consequent chapter.

THE MARXIAN CIRCLE: LABOR, TECHNOLOGY, AND NATURE

The starting point for Marx's materialist inversion of Hegel is Feuerbach's contention that "we know the man by the object," which is "*his manifested nature, his true objective ego.*" Developing this methodological premise, Marx departs from what he considers a contemplative form of Feuerbach's anthropology, replacing the *ideal* (and, as such, one-sided) activity of socialized self-consciousness with the activity of *productive labor*; labor is both *material*, sensuous activity, in which man "no longer reproduces himself merely intellectually, as in consciousness, but actively and in a real sense," and *objective* activity, allowing man to see "his own reflection in a world which he has constructed."[4]

Marx underlines an intrinsic unity between objective human essence, objective activity, and its products: "An objective being <u>acts objectively</u>, and it would not act objectively <u>if objectivity were not part of its essential being</u>... it does not descend from its 'pure activity' to the **creation of objects**; its **objective** product simply <u>confirms</u> its **objective** activity...." This unity constitutes man as a *being for himself* and thus a universal (species) being: "**man is not merely a natural being; he is a human** natural being. **He is a being for himself, and therefore a species-being.**"[5] Marx points out that in productive labor, man can enter into the *for himself* relationship with both nature and other individuals; it is the *for himself* mode of labor that qualifies it as *objective* activity, 'confirming' the objectivity of man as a universal social, natural, and, hence, a *total* being.

The human *for himself*, "active orientation" to the world is made possible due to the *self-referential* character of labor. In the *Paris Manuscripts* of 1844, Marx characterizes man as *essentially self-referring*, with all his powers having a "quality of self-reference": "man is self-referring... Every one of his faculties has this quality of self-reference... [and]... the real, active orientation of man to himself as species-being (i.e., a human being) is possible so far as he really brings forth all his species-powers (which is only possible through the co-operative endeavors of mankind and as an outcome of history and treats these powers as objects."[6]

The self-referential pattern of labor essentially recapitulates the Feuerbachian dialectic of socialized consciousness, where men realize their universal nature through mutual self-recognition: the *outward projection* of oneself onto another individual and the application of the consequently *objectified* (through another) image to themselves. While for Feuerbach, the activity of the reciprocal self-recognition is confined to self-consciousness, the material character of labor allegedly allows man to establish *real* ('objective') identity with other men as well as nature:

---

[4] Marx, K. (1971). *Economic and Philosophical Manuscripts* In Fromm, E. *Marx' Concept of Man: with a Translation from Marx's Economic and Philosophical Manuscripts* (p. 102) (T. B. Bottomore. Trans.). Frederick Ungar Pub.

[5] Marx, K. (1971). *Paris Manuscripts*, p. 181.

[6] Marx, K. (1971). *Paris Manuscripts*, pp. 178–179.

"It is just in his work upon the objective world that man really proves himself as a **species-being**. This production is his active species-life. By means of it nature appears as **his** work and his reality. The object of labor is, therefore, the **objectification of man's species life....**"[7]

Though Marx considers self-consciousness as a byproduct of the self-referential nature of man, he does not delve into the material aspects or genesis of self-reference. By itself, self-reference could be viewed only as a particular form of the negative feedback mechanism exhibited by any self-regulating system and shared by men and animals. To the extent that self-reference secures the *universality* of man (as a *being for himself*), the relation between self-reference and self-consciousness is reversed: Marx implicitly assumes that man is a *self-referential* being because he is a *self-conscious* being.

Most importantly, Marx's contention of the self-referential nature of man is essential for his conception of technology. Marx underlines that labor "must not be considered simply as being the production of the physical existence of the individuals." It is due to the mediating role of *technology* that man produces and reproduces himself as a universal being and establishes a universal relationship with nature. By means of technology, as an embodiment of essential human powers, "we gain an understanding of the *human* essence of nature or the *natural* essence of man."[8] Marx reiterates that technology represents human "manifested nature," so that the objective essence of men "coincides with their [mode of] production."[9]

In his account of technology, Marx follows Hegel, who argues that human subjectivity in its teleological capacity contains "an essential striving and impulse to posit itself externally." To accomplish that purpose, "it posits itself in a *mediate* connection with the object, and between itself and this object *inserts* another object";[10] the latter serves as a *means* employed by the subject to achieve his purposes (satisfy needs) in interaction with an external environment. It means that "the rationality in the purpose maintains itself as such maintaining itself in this *external order*, and precisely *through* this externality." Hegel continues, "to this extent the *means* are higher than the *finite purposes* of the *external* purposiveness." The practical import of this dialectic of means and ends is that the

---

[7] Marx-Engels. (2010). *Paris Manuscripts*, v. 3, p. 277.
[8] Marx-Engels. (2010). *Paris Manuscripts*, v. 3, p. 303.
[9] Marx, K. (1978). *The German Ideology* In Karl Marx *Selected Writings*, pp. 160–161.
[10] Hegel, G. W. F. (2010). *The Science of* Logic, p. 663.

means, represented by *tools* (such as the plough) and instruments, are "higher" than perishable products of labor intended for the satisfaction of human needs. Hegel emphasizes that "it is in their tools that human beings possess power over external world"; accordingly, "the *plough* is more honorable than are immediately the enjoyments which it procures and which are the purposes."[11] Underlying the continuity between the Hegelian and Marxian dialectic, Lenin, in the *Philosophical Notebooks*, refers to these ideas as "the elements of historical materialism."[12]

Utilizing this conceptual apparatus in his dialectic of technology, Marx says that "**Everyday material industry** shows us, in the form of **sensuous useful objects**, in the alienated form, the **essential human faculties** transformed into objects... [and] conceived as the **exoteric manifestation** of the essential human **faculties**." Tucker summarizes Marx's historical dialectic of technology-based labor in the following way: "Machines, factories, etc., are materialized faculties of generic man's self-expression in productive activity. They are physical extensions and enlargements of the hands, ears, eyes and brains of the species. Taking his cue from Hegel, Marx says that the history of production is an Entäusserungsgeschichte, a history of man's own self-externalization."[13] The logic of self-externalization could be formalized and explicated in the context of von Neumann's theory of self-reproducing automata, where every automaton's 'physical organization' would play a role similar to genetic code in biological *onto-genesis*. In such a framework, *techno-genesis* could be interpreted as a collective self-externalization of humans engaged in developing a distributed technological infrastructure of their 'objectified' and amplified physical organization ('essential human faculties').[14] As a matter of fact, such formalism could reasonably represent the whole class of self-referential models, including the ontogenesis, Hegelian dialectic of the

---

[11] Hegel, G. W. F. (2010). *The Science of Logic*, pp. 657–653.
[12] Ленин, В. И. *Философские Тетради;* Ленин, В. И. *Полное Собрание Сочинений* (Collected Works), т. 29. стр. 171.
[13] Tucker, R. C. (1961). *Philosophy and Myth in Karl Marx* (pp. 130–131). Cambridge University Press.
[14] Neumann, J. Von. (1966). *Theory of Self-Reproducing Automata*. University of Illinois Press. A. Bogdanov offers an alternative formalization of Hegelian dialectic in the 'general theory of organization,' often viewed as a precursor of *cybernetics* and the *general systems theory*. We will see in the next chapter that such formalism could be traced to Herbert Spenser's theory of universal evolution.

'substance working within itself' and Marx's model of technological self-externalization.

The dialectic of technology, a key element of Marx's conception of man as a total being, is at the foundation of his humanism. He maintains that due to the mediating role of industry in human interaction with nature, "the **human** essence of nature and the **natural** essence of man [all underlining is by YK] can also be understood." This correspondence between 'subjective' and 'objective' nature is based on the historical dialectic of technology: "Nature, as it develops in human history, in the act of genesis of human society, is the **actual** nature of man; thus nature, as it develops through industry, though in an **alienated** form, is truly **anthropological** nature."[15] While "animals produce only themselves," in the way determined by nature, man, through the use of technology, "reproduces the whole of nature," without any alien addition (in Engels' words).[16] Marx's anthropocentric ontology of nature is explicated in the *Paris Manuscripts*: "neither objective nature nor subjective nature is directly presented in a form adequate to the **human** being." He drives this point home by bluntly stating that "nature too, taken abstractly, for itself, and rigidly separated from man, is nothing, for man."[17] In a spirit of transcendental tradition, the objectivity of nature emerges *for man* in the course of human activity, while the ontological status of nature *in itself* remains (phenomenologically) 'bracketed.'

While for Kant, nature is *indirectly* presented to man through the activity of pure reason, for Marx, nature emerges, in the *indirect* "form adequate for human being," through the material, practical activity mediated (and restricted in its scope) by technological externalization of human physicality: "The first premise of all human history is, of course, the existence of living human individuals. Thus the first fact to be established is the physical organization of these individuals and their consequent relation to the rest of nature...."[18] In its "consequent" form, nature emerges as an *extended body* of man: "a human being's relation to his natural conditions of production" means nothing else but "relations to them as **natural**

---

[15] Marx, K. (1971). *Paris Manuscripts*, p. 136.
[16] Marx, K. (1971). *Paris Manuscripts*, p. 102.
[17] Marx, K. (1971). *Paris Manuscripts*, p. 193.
[18] Marx continues: men "begin to distinguish themselves from animals as soon as they begin to produce their means of subsistence, a step which is conditioned by their physical organization... What they are, therefore, coincides with their production, both with what they produce and with **how** they produce." Marx, K. (1978). *The German Ideology*, p. 160.

presuppositions of his self, which [i.e. nature] only form, so to speak, his extended body."[19] In this framework, the *isomorphic* relationship between man and nature constitutes both the objectivity of nature and the objectivity (universality) of man as a natural being. Marx emphasizes that, for man, genuine unity with nature, as a human extended body, is possible only through the emancipation of labor relations. Moreover, the universality of social relations is inseparable from the universal mode of human attitude towards nature.

Marx's *anthropocentric ontology* of nature clearly distinguishes his philosophy from the preceding forms of 'contemplative' materialism. For Marx, 'nature in itself' is "nothing" to the extent that it is instrumentally bracketed, and, in his 'transcendental materialism,' this is the most foundational form of bracketing.

Marx's instrumental phenomenology of labor seemingly resonates with Heidegger's ontology of technological enframing. For Heidegger, physics "sets nature up to exhibit itself as a coherence of forces calculable in advance, it orders its experiments precisely for the purpose of asking whether and how nature reports itself when set up in this way." In this context, the progress of physics, its scope and extent, depends on technological advancement. Heidegger distinguishes his position from Kantian purely theoretical apriorism by emphasizing that "modern physics, as experimental, is dependent on technological apparatus."[20] In this context, he underlines that scientific "asking" is inseparable from technological *enframing* [*Ge-stell*] turning nature (and man himself) into the standing-reserve. While for Heidegger, enframing is a path toward losing the potentialities of freedom rooted in human being-in-the-world (*Dasein*), Marx conceives technological self-externalization as a means of liberation.

Though the dialectic of technology and its impact on economic, political, and ideological superstructure lie at the foundation of Marx's concept of history, he never developed a comprehensive theory of technological progress. In *Grundrisse*, he says that "political economy is not technology...

---

[19] Marx, K. (1973). *Grundrisse, Foundations of the Critique of Political Economy* (pp. 415–416) (M. Nicolaus. Trans.). Penguin. This position is sharply different from the Fichtean understanding of nature as an alienated pseudo-material screen for the infinitely unfolding subjective activity.

[20] Heidegger, M. (1993). "The Question Concerning Technology" In Heidegger, M. *Basic Writings* (pp. 319–320, 326). HarperCollins Pub. We address Heidegger's philosophy of technology in Chap. 5.

[which is] to be developed elsewhere (later)."[21] Moreover, at the outset of *The German Ideology*, Marx says that he does not want to "go either into the actual physical nature of man, or into the natural conditions" of human existence. At the same time, in the *Capital*, he attempts to link Nature's Technology to human technology and points to the possible analogy between Darwin's biological evolution and technological progress:

> Darwin has interested us in the history of Nature's Technology, i.e., in the formation of the organs of plants and animals, which organs serve as instruments of production for sustaining life. Does not the history of the productive organs of man, of organs that are the material basis of all social organization, deserve equal attention?... Technology discloses man's mode of dealing with Nature, the process of production by which, he sustains his life, and thereby also lays bare the mode of formation of his social relations, and of the mental conceptions that flow from them."[22]

Unfortunately, Marx left this promising conceptual path largely unexplored. Though not pursued by Marx, this line of thought is explored within the so-called *organ projection theory* advanced by Ernst Kapp (1808–1896) and developed by Ludwig Noire (1829–1889), A. Galen and others.[23] Reinforcing conceptual similarities between the logic of organ projection and Marx's dialectic of labor, Bogdanov, one of the most consistent interpreters of Marx's s anthropocentric ontology, refers to Noire's version of organ projection theory as a natural complement to Marxism.

Bogdanov's assessment is not unfounded, considering that Kapp refers to Feuerbach as one of his precursors and employs a dialectical pattern of self-externalization to explain technological evolution as a 'projection' and the consequent 'objectification' of human physicality onto nature. Methodologically, Kapp characterizes his theory as 'actual empiricism,' meaning "that everything emanating from the human being is in fact his

---

[21] Marx, K. (1973). *Grundrisse*, p. 86.
[22] Marx, K. (1967). *The Capital*, vol. 1. In Marx & Engels (2010). *Collected Works* (Vol. 35, p. 375, n.2). Lawrence and Wishart.
[23] Kapp, E. (1877). *Gründlinien einer Philosophie der Technik—Zur Entstehungsgeschichte der Cultur aus neuen Gesichtspunkten*, Braunschweig; translated as Kapp, E. (2018). *Elements of a Philosophy of Technology: On the Evolutionary History of Culture* (3rd ed.). University of Minnesota Press. Gehlen, A. (1961). *Anthropologische Forschung*, Reibek. Noire, L. (1880). *Das Werkzeug und seine Bedeutung für die Entwicklungsgeschichte der Menschheit*. (1968 reprint is available).

own nature—human nature—which he disperses throughout the world and which is displayed before him as the system of human needs. The system of needs—the substance transfigured via organ projection into tools and equipment…—is the outside world… In this outside world, the human being becomes self-conscious, truly rediscovering himself in it and learning to know and comprehend himself." Linking the evolution of human instruments to 'nature's technology,' Kapp points to human "original disposition to create in his tools the artificial organs with which to comprehend the stimuli affecting him from without." Such a disposition lies in "the depth of the unconscious, indeed to be no more than the consciousness first unbound in primitive tools." Gradually, this activity involves more and more complex devices and machines, progressively substituting the instrumental functions of the human body (starting with a hand).[24] While Kapp appeals to Feuerbachian idea of human self-objectification, Noire construes a transition from the 'depth of unconscious' to its instrumental manifestation in terms of Kantian transcendental construction (and consequent objectification), which is rooted in the efficacy of 'productive imagination.'

Overall, the organ projection theory shares with Marx's dialectical anthropocentrism the view of technology as an externalized form of human physical organization ('species powers'). Methodologically, the logic of projection and objectification combines the principles of Kantian constructivism and the Feuerbachian dialectic of self-consciousness. At the same time, this theory utilizes philological, archaeological, and physiological data to demonstrate the evolutionary links between the functions and structure of the human body (and organic life in general) and human technology.

By contrast, Marx's dialectic of labor remains predominantly within the conceptual scope of German transcendental philosophy. On this background, Marx's substitution of the abstract notion of 'universality' with the allegedly concrete notion of 'objectivity' only obscures the umbilical cord between his dialectic of labor and the Hegelian dialectic of the 'kind' (and, by extension, the abstract Feuerbachian dialectic of social self-consciousness). In this framework, infinite qualitative technological progress—'genetically' predetermined in its makeup, scope, and extent by an *isomorphic* relationship between human physicality and nature—is

---

[24] Kapp, E. (2018). *Elements of a Philosophy of Technology*, p. 92.

inconsistent with the close-ended dialectic of technology as an **"exoteric manifestation of the essential human faculties."**

In the last analysis, Marx's conception of labor as a unity of the universal (formal) and material modes of human activity undermines the epistemological foundations of historical materialism, while his dialectic of labor reveals the inconsistency between the open-ended conception of history and the assumption of infinite technological growth.

Infinite Growth and Dialectic of Self-Externalization

The ideologeme of infinite technological growth lies at the foundation of the Marxist providential Idea of Progress. In the 'Draft of a Communist Confession of Faith,' a document discussed at the first congress of the Communist League in 1847, Engels explicitly refers to the unlimited growth of technology and material resources:

- How do you wish to achieve the aim [of the communists]?
- By the elimination of private property and its replacement by community of property.
- On what do you base your community of property?
- Firstly, on the mass of productive forces and means of subsistence resulting from the development of industry, agriculture, trade and colonization, and on the <u>infinite possibility</u> inherent in machinery, chemical and other resources of their <u>unlimited growth</u>.[25]

This draft was later used by Engels for another program document of the League, 'The Principles of Communism,' where the postulate of unlimited growth remains in place: "What are the further results of the industrial revolution?—In the steam-engine and the other machines large-scale industry created the means of <u>increasing industrial production</u> in a short time and at slight expense <u>to an unlimited extent</u>."[26] In turn, this

---

[25] Engels, F. (1976). "Draft of a Communist Confession of Faith" In Marx, K. & Engels, F. *Collected Works* (Vol. 6, p. 96, translation modified). International Publishers. This document was found in 1968 in the archives of the active member of the Communist League, J. F. Martens.

[26] Engels, F. (1976). "Principles of Communism" In Marx, K. & Engels, F. *Collected Works* (Vol. 6, p. 341).

work inspired the famous *Communist Manifesto* (1848), "published as the platform of the Communist League."[27] In addition, in the *German Ideology* the founders emphasize the necessity of unceasing and global ("world-historical") development of productive forces for the success of communist struggle: the "development of productive forces (which itself implies the actual empirical existence of men in their world-historical, instead of local, being) is an absolutely necessary practical premise because without it want is merely made general, and with *want* the struggle for necessities would begin again and all the old filthy business would necessarily be restored."[28]

The idea that technological expansion constitutes a key driver and a precondition of social progress can also be found in 'mature' Marx's works, such as *Grundrisse* and *Capital*. He emphasizes that under both capitalism and socialism, consumption and production are intimately intertwined, supporting each other in the infinite self-propelled cycle, and elaborates on this dual relationship, explaining that "with his [man's] development this realm of physical necessity expands as a result of his wants; but, at the same time, the forces of production which satisfy these wants also increase."[29] In the *Critique of Gotha Programme* (1875), Marx further underlines the importance of material production for the advancement of socialist society, stating that "after the productive forces have also increased... [and] all the springs of co-operative wealth flow more abundantly—only then... society inscribe on its banners: From each according to his ability, to each according to his needs!"[30]

At the same time, the stage of the 'abundant' distribution of wealth still belongs to the 'realm of necessity' ('prehistory'), where "freedom... can only consist in socialized man, the associated producers, rationally regulating their interchange with Nature, bringing it under their common control." The ideal society—*communism*—lies beyond the 'realm of necessity' and constitutes "the true realm of freedom, which, however, can blossom forth only with this realm of necessity as its basis," with machines gradually replacing human labor and providing leisure for the creative

---

[27] Preface to the English edition of the *Communist Manifesto* (1888).
[28] Marx, K. and Engels, F. (1998). *The German Ideology* (p. 54). Prometheus Pub.
[29] Marx, K. (1973). *Grundrisse* (p. 704) (M. Nicolaus. Trans.). Vintage Books. Marx. K. *Capital*. (1967) (Vol. 3, p. 820). International Publishers.
[30] Marx, K. (1970). *Critique of the Gotha Programme* In Marx, K. & Engels, F. *Selected Works* (Vol. 3, p. 13). Progress Pub.

development of unspecified human 'species powers.'³¹ In such a society, the truly humane relations within the community imply the truly humane relations of the community towards nature, so that in an emancipated (communist) society, the "conflict of man's individual, material existence with his species-existence… [is] superseded."³² Thus, in the Marxian social utopia, infinite self-propelled technological expansion is necessary for both historical transition to and sustainable advancement of the communist 'realm of freedom.'

Marx's followers have tried once and again to justify the infinite potential of scientific and technological appropriation of the world. Their arguments are typically based on the assumption of an 'irreducible' ontological gap between man and nature or/and on the claim of the unlimited cognitive and instrumental potential of man as a *universal being*. These arguments do not take into account that Marx's *humanism* represents a complicated balancing act between materialism and idealism. This position is misunderstood by his interpreters from left to right, who assume that Marx simply combines Hegelian dialectic with the concept of nature 'in itself.'

The typical line of such argument is presented by Arthur: "It is the irreducible distinction between man and the objective basis of his activity, however intermediated through labor and industry, that allows us to grasp the dialectic of human practice as historical and open-ended."³³ In the words of another Marxist scholar, man is able "to turn a principally unlimited scope of natural laws, regularities into the principles of his own actions and to transform his progressively expanding environment to an ever-increasing degree."³⁴ This argument further advances the idea of the *universality of labor* as "the essential activity of man" who "… is essentially **universal natural being** in a sense that he is **potentially** able to turn any object of nature into the subject matter of his wants and activity…."³⁵ We find a more nuanced approach to Marx's ontology in Habermas' reference to "the paradoxical consequences of taking Fichte's philosophy of the ego and undermining it with materialism. Here, the appropriating subject

---

[31] Marx, K. (1967). *Capital*, v. 3, p. 595. Already Aristotle, recognizing pragmatic value of technology, observes in the *Politics* that machines could replace slave labor.

[32] Marx, K. (1978). "On the Jewish Question" In Karl Marx *Selected Writings*, p. 62.

[33] Arthur, C. J. (1986). *Dialectics of Labor* (p. 133). Basil Blackwell.

[34] Markus, G. (1978). *Marxism and Anthropology* (pp. 12–13). The Netherlands: Van Gorcum Assen.

[35] Marcus. (1978), p. 12.

confronts in the non-ego not just a product of the ego but rather some portion of the contingency of nature."[36] Habermas refers to the assumingly "paradoxical" opposition between the Fichtean concept of the ego as a subject of transcendental construction and the concept of nature in its contingent ('in itself') existence.

The claims of "irreducible distinction between man and the objective basis of his activity" and of nature as a source of "a principally unlimited scope of natural laws" reflect a contemplative concept of nature as an independent *in itself* "objective basis" of human activity, effectively reducing Marx's anthropocentric ontology to the 'orthodox' philosophy of *dialectical materialism*, often supported by the unwarranted assumption of a radical break between the 'early' and 'mature' Marx.[37] The critics do not take into consideration that, for Marx, labor represents an *objective mode* of human activity *commensurate* with the objectivity of nature as it is being revealed in human praxis. Marx emphasizes that for the philosophy of true *humanism*, nature *emerges* for man in its objective mode—as "**his** work and his reality"—through the *for itself* mode of instrumental (labor) activity. For him, "neither objective nature nor subjective nature is <u>directly</u> presented in a form adequate to the **human** being... and rigidly separated from man, is nothing, for man."[38]

On this account, in Marx's ontology, nature does not 'confront' man but is *coextensive* with him as a *universal being*. In the course of history, it is being revealed to man as his *extended body*, allegedly resolving one of the principal 'conflicts' of human alienated existence. As an extended body of man, nature represents neither the 'thing in itself,' as an inexhaustible resource for human theoretical and instrumental activity, nor the subjective Fichtean 'horizon' of infinite activity of the (absolute) I. As a contingent product of (inexhaustible in itself) nature, a human being would not be able to enter into a *universal* relationship with the 'whole of nature'; and, as a purely spiritual being, man would not be able to enter into any *material* relationship with nature; in either case, man would not have "a

---

[36] Habermas, J. (1971). *Knowledge and Human Interests* (p. 44) (J. J. Shapiro. Trans.). Beacon Press.

[37] Kolakowski points to the "clear difference between... Engels' dialectic of nature and the dominant anthropocentrism of Marx's view." Kolakowski, L. (1978). *Main Currents of Marxism* (Vol. 1, p. 402). Clarendon Press. In "The Young and the Old Marx," Wartofsky emphasizes continuity of Marx's thought in *Marx and the Western World*. (1967). (p. 39) (N. V. Lobkowitz. Ed.). University of Notre Dame Press.

[38] Marx, K. (1971). *Paris Manuscripts*, p. 193.

principally unlimited scope of natural laws" at his disposal and would not be "able to turn any object of nature into the subject matter of his wants."

Most importantly, the assumed universality of labor does not imply the *actual infinity* of "an ever-widening, principally unlimited range of natural regularities." It is the *isomorphism* between subjective and objective nature that constitutes the *universality* of man as a *natural* and, as such, a truly *human* being. From this perspective, the assumptions of the infinity of nature or human 'inexhaustible' cognitive powers are redundant for the conception of human universality. As we argue in the next chapter, the universality of science also does not imply the infinite cognitive or instrumental potential of human appropriation of nature or its infinite structural complexity, and the postulate of the infinite expansion of (techno)science represents nothing more than a regulative idea of reason. In this framework, the human physical organization serves as an 'a priori' matrix, *schemata* for the historical 'techno-genesis,' and natural laws express nothing but the *invariants* of human experience.

Some Marxist scholars find an epistemological justification for infinite 'opening out' in the distinction between *objectification* and *alienation*. The former is viewed as a positive phenomenon, while the latter is negative. Sartre says, in this respect:

> For Marx, indeed, Hegel has confused objectification, the simple externalization of man in the universe, with alienation which turns his externalization back against him. Taken by itself—Marx emphasizes this again and again — objectification would be an opening out; it would allow man, who produces and reproduces his life without ceasing and who transforms himself by changing nature, to 'contemplate himself in a world which he has created'[39]

Sartre addresses a humanistic perspective on objectification as human self-affirmation in the world. This interpretation indeed resonates with Marx's objective of "emancipation of all human senses and qualities," allowing man "to contemplate himself" in his unity with nature. However, this type of self-affirmation comes into conflict with a Promethean view of 'unceasing' self-externalization, which is necessary, according to Marx, for the transition to and sustainable development of socialism.[40] In the

---

[39] Sartre, J. P. (1963). *The Problems of Method* (p. 13) London: Methuen & Co Ltd.
[40] Marx. (2010). *Paris Manuscripts*, vol. 3, pp. 291–300.

ever-expanding cycle of industrial production and multiplication of human wants, man inevitably gets entangled in the utilitarian net of "crude, practical need," effectively turning him into a universal consumer being instead of a universal species being.[41] It is in this context that Spengler's assessment of the Marxist ideal of socialist society as reverse capitalism appears justifiable.

Arguing against the productivist reading of Marx, Saito (2022) claims that, in his late period, Marx "consciously discarded historical materialism" and adopted a position of *degrowth communism*.[42] He justifies this position by first referring to Marx's concerns with the environmental impact of capitalism in the notebooks of 1860th; and second by highlighting Marx's shift from strict Eurocentrism to the broader historical outlook. In the last respect, Saito points to the Preface to the Russian edition of the *Communist Manifesto*, where the Founders refer to the importance of the Russian agricultural commune for the revolutionary movement.

However, true to the spirit of historical materialism, Marx and Engels specifically stipulate that the victorious Russian revolution would have to be accompanied by the simultaneous proletarian revolution(s) in the West. Thus, they reaffirm the leading role of the proletariat in the social liberation movement, which was already highlighted in Marx's early writings and maintained in the 'mature' period. In the 1844 *Paris Manuscripts*, Marx emphasizes that the source of social alienation is the "life of private property—labor and conversion into capital" and concludes that "the human being had to be reduced to this absolute poverty in order that he might yield his inner wealth to the outer world"; the class reduced to 'absolute poverty' and capable of discovering the 'inner wealth' and the universal nature of humanity is the proletariat. In this context, we cannot agree with Saito that Marx's attention to the 'archaic' mode of production, in general, and the Russian agricultural commune, in particular, constituted a breakup with historical materialism. In fact, at the time, Russia did not have the mature political base for a successful socialist revolution. The Russian socialists Marx and Engels corresponded with—most notably, Petr Lavrov, Vera Zasulich, and Nikolai Mikhaylovsky—belonged to the intelligentsia and had little connection with the socioeconomic base of the gradually brewing revolution.

---

[41] Marx. (2010). *Paris Manuscripts*, v. 3, pp. 300, 302.
[42] Saito, K. (2023). *Marx in the Anthropocene* (p. 182). Cambridge University Press.

To support the view of Marx as a 'degrowth communist,' Saito employs a distinction between *natural* vs. *social* scarcity. While the first is characteristic of any mode of production, capitalism and socialism included, the second would be assumingly eliminated in a socialist/communist society. The elimination of social scarcity, argues Saito, would lead to reducing human wants to the 'natural' level, allowing the maintenance of social harmony under the conditions of zero growth; that is what, in a nutshell, the theory of degrowth is.

On the surface, this position is consistent with Marx's early call for the emancipation of human senses from the sense of having, a view inherently linked to the assessment of the proletariat as a universal class capable of acting 'under the veil of ignorance.' However, Saito seems to ignore that in Marx's ontology of labor, nature is being *revealed for man* in the course of technological self-externalization while *in itself* it is 'nothing.' On this account, the 'natural' scarcity 'in itself' is also nothing for man and represents a *social construct* dependent on the level of production forces, including the technological base. As recognized by Marx, man "transforms himself by changing nature" through historically unfolding labor activities, with new 'natural' wants constantly generated by technological expansion. Moreover, *infinite technological progress* would generate an *infinite multiplication* (and means of satisfaction) of *natural* wants, thus completely relativizing the relationship between natural and social scarcity. In this context Marx remarks that continuously growing wants would be satisfied with the expansion of production. What we could assume is that with the elimination of the capitalist production system, the expansion of human wants would be reduced to a sustainable level consistent with the technological growth (but not 'degrowth') in a planned socialist economy; the alternative, in Marx's view, is a return to the 'filthy business' of socioeconomic suffering and struggle.

The phenomenon of expanding human wants can be illustrated by the developments in the Soviet Union, starting with the October 1917 Revolution. Martov, one of the leading figures in the early twentieth-century Russian revolutionary movement and an opponent of Bolshevism, pointed to the strengthening, at the time, phenomenon of 'consumer communism' and called for the 'psychological emancipation' of the proletariat.⁴³ The principle of psychological emancipation became an essential ingredient of Bolshevik ideology.

Yet, the program of psychological emancipation failed already in the early 1920th, foreshadowing the collapse of the promised social paradise. The persistent spirit of consumerism was only reinforced by the Bolshevik's New Economic Policy (NEP), legalizing the elements of the market economy to save the country from hunger and devastation caused by the civil war and the policy of wide-ranging expropriations. Viewed by many as a betrayal of the ideal of elimination of the sense of having, and amidst the widespread corruption, NEP came to a premature end in the political struggle of the late 1920th, giving way to the Stalinist draconian 'emancipation,' directed at molding 'new man' (the 'last man' in Fukuyama's language) of the future. Through the consequent sixty-year journey in the dictatorial desert, the party vainly tried to instill the desired second nature into its ranks and the 'proletarian' masses.

This problem revealed the fundamental flaw of the Marxist view of man as a universal being and, by extension, the proletariat as a universal class. Indeed, the NEP was an answer to an early realization that the widespread poverty and social revolution did not entail the abrupt elimination of the 'sense of having' and only undermined (socially and psychologically) a viable management system. The collapse of the Soviet Union is a witness to the gaping chasm between Marx's idea of man as a universal being and its historical implementation. At the core of the problem is a perennial dilemma of freedom and necessity, the universal human nature vs. human psychology, transparent already in the Cartesian dualism of man as an ideal thinking substance in the material body machine and the Kantian dichotomy between good will and the (evil) nature of a concrete individual.

The unresolved conflict between the elimination of the sense of having vs. the persistent psychology of consumerism remained a characteristic feature of the Soviet way of life, becoming a major factor contributing to the eventual collapse of the Soviet system in 1991. The stunning incompetence of the 'elites,' amplified by consumerism, growing corruption, and criminality throughout the 'wild' 1990th, de facto transformed Russia into a mafia state. Finally, after thirty years of failed attempts at integration into the Western economy, turning Russia into a cheap world resource

---

⁴³ Мартов, Ю. О. (1923). *Мировой Большевизм*. Berlin: Iskra (Martov. *The World Bolshevism*). This suggestion could be viewed as instilling into people the *second nature*, in Aristotelian sense.

base, the local power elites abruptly changed the course and again adopted policies of forced psychological emancipation. On the heels of the 2022 Russian-Ukrainian war, the geopolitical confrontation triggered the revamping of public education, strengthening censorship, persecuting the allegedly 'foreign agents,' and pressuring the opponents of the emerging police state into emigration. These practices remind the balancing act between the persecution and "preventive mercy" of the Russian cultural elite by the Bolsheviks a hundred years earlier. As Trotsky said at the time, "We have nothing to execute them for, but we cannot tolerate them here."[44]

The recent reversal of the brief surge of (pseudo)liberal freedoms in Russia resembles the regression of the Greek Enlightenment at Athens about 2500 years earlier, during the wartime crises marked by the prosecution of Socrates, Anaxagoras, and Protagoras, among other intellectuals. The breakup with enlightened rationalism is reflected in Plato's last work, *Laws*, where he declares that no person "be ever suffered to live without an officer set over them, and no soul of man to learn the trick of doing one single thing of its own sole motion." In modern Russia, the new state mythology is being created by merging the narrative of the Russian Empire with the spirit of late Stalinism and the neo-Hegelian worship of the State. History indeed repeats itself in a peculiar way.[45]

To summarize, even if objectification would indeed be different from alienation (as Sartre suggests) under the socialist system of production, whether "the simple externalization of man in the universe" can proceed 'without ceasing' is a different matter. As we have tried to show, objectification 'without ceasing' is incompatible with the Hegelian structure of Marx's dialectical apparatus, leaving no room for the infinite 'opening out.' In other words, the historical dialectic of labor, loaded with the

---

[44] The interesting observations on the 1920th politics are in Бажанов, Б. (1992). *Воспоминания Бывшего Секретаря Сталина* (Bazhanov, B. G. *The Memoirs of Former Stalin Secretary*). Ст. Петербург: Всемирное Слово (1930 original publication). In the words of Prince Trubetskoy, one of those expelled from Russia by the Bolsheviks in 1922, "The Soviet power, as wrote Trotsky, defeated all its internal enemies… But… the remaining internal enemies will definitely join the external enemies of the Republic if it comes (something not excluded in the future) to a direct confrontation with them. Certainly, our internal enemies will still present no threat to us; nevertheless, we will be forced to destroy them. Thus, preventive mercy… impels us in advance to throw the rest of our internal enemies out of the country." Трубецкой, С. Е. (1991). *Минувшее* (The Past). ДЭМ.

[45] See Dodds, E. R. (1951). *The Greeks and the Irrational* (Chap. 6–7). University of California Press. *The Laws of Plato*. (1934) (Book XII, 942) (A. E. Taylor. Trans.). J. M. Dent & Sons Ltd.

notions of self-reference, objectification, alienation, negation, etc., is inconsistent with a widely accepted ideologeme of infinite technological growth. Furthermore, as the inconsistency between the circular dialectic of labor and the postulate of infinite technological progress undermines the 'Promethean spirit' of historical materialism, the inconsistency between the objective of eliminating the sense of having and the assumption of unlimited growth and satisfaction of the growing human wants undermines its value structure.

In the rest of the chapter, we briefly explore a controversy between the early twentieth-century orthodox and revisionist interpretations of Marx. We contrast the philosophy of dialectical materialism with the distinctly anthropocentric reading of Marx's philosophy developed by Lenin's one-time ally, A. Bogdanov, opening the way to the discussion of the limits to science and technology issue in the next chapter.

Marxist Orthodoxy and Beyond

In the words of Kolakowski, Marx's "influence extended beyond the immediate circle of the faithful, to historians, economists, and sociologists who did not profess Marxism as a whole but adopted particular Marxist ideas and categories."[46] This influence has been so broad that labeling one as a Marxist has become almost meaningless, and even Marx once remarked (assumingly referring to French Marxists) that he was not a Marxist.[47] As a result, already during Marx's lifetime, his self-proclaimed followers fragmented into numerous, loosely related groups and the distinction between the orthodox and non-orthodox interpretations of Marx became more a matter of political power struggle rather than an outcome of a balanced debate and a thorough analysis of Marx's teaching. For example, in the fierce argument over the October 1917 Russian Revolution, Karl Kautsky (1854–1938), one of the 'orthodox' leaders of the *Second International*, was attacked by the 'grandmaster' of the movement V. I. Lenin (1870–1924) and labeled as a 'renegade.' In his militant style, Lenin remarks that Kautsky's book on the dictatorship of the proletariat is "hundred times more shameful, more outrageous, more renegade than the famous *Premises of Socialism* by Bernstein," who became a common target

---

[46] Kolakowski, L. (1978). *Main Currents of Marxism*, vol. 2
[47] As reported by Engels in a letter to Konrad Schmidt, 08.05.1890. In Marx, K. & Engels, F. (1965). *Collected Works* (Vol. 37, p. 370) (2nd ed.). Moscow: Political Literature Pub.

of critique for the broad spectrum of 'orthodox' Marx's followers (including Kautsky).[48]

The key issue in the disagreement between Lenin and Kautsky is whether the dictatorship of the proletariat should be considered as a form of government (as Lenin insists, claiming that the Soviets successfully implemented this principle) or as a guiding idea for a democratic majority rule (as Kautsky argues). This controversy pertains to the very core of Marx's theory of humane society and the conception of man as a universal being. While in Marx's social theory, universality is, *in principle*, assigned to a socialized individual, *in practice*, it is transferred to a particular socioeconomic group—*the proletariat*—as a *universal* class. The increasingly public—and therefore increasingly *universal*—way of production allegedly enables the proletariat to make free and fair political and social choices as if 'under the veil of ignorance' (using Rawls' expression); in this way, the proletariat becomes a vehicle and a 'vanguard' of political struggle against the dehumanizing socioeconomic conditions of capitalism.[49] What appears in Rawls' 'Cartesian meditations' on social justice as a matter of mental experiment constitutes an essential element (and a prescription for action) of Marx's socioeconomic theory.

While the concept of the proletariat allows Marx to present the driving force of social progress as (in Hegel's terms) a 'concrete historical universal,' it undermines the vitally important concept of man as a universal (species) being (or *being for himself*, as Marx calls it). The problem becomes apparent when we consider that in the socialist society, the proletariat loses its status as the impoverished class destined to play a leading role in social progress. A case in point is that immediately after the 1917 Russian Revolution, Lenin and his followers realized that the founders of Marxism contributed next to nothing to the principles of a planned socialist economy. While initially Lenin naively expected that the transition from the capitalist to socialized economy required only to 'change the system of

---

[48] Lenin, V. I. (1969). "Renegade Kautsky and Proletarian Revolution." In Lenin, V. I. *Collected Works* (Vol. 37, p. 101) (5th ed.). Moscow: Political Literature Pub. The article is originally written in 1918, soon after the Bolsheviks' overthrow of the Provisional Government and the consequent replacement of the parliamentary system by the Soviets (in theory, representing the 'proletariat'). Kautsky, K. (2010). *The Social Revolution: Essays on the Socialist Reform and Revolution*. Red and Black Pub. (Reprint of the edition by Charles Kerr & Co.).See also Kautsky. K. (1918). *Die Diktatur des Proletariats, Wien*. (This book is referred to by Lenin).

[49] Rawls, (1999). *A Theory of Justice*. Belknap Press.

accounting,' it turned out that the *public way of ownership* does not automatically entail the viable *public way of management* (by proletariat per se). The management functions could not be implemented through the popular vote or effectively handled by the Soviets. What the leaders of the Revolution finally realized is that Marx, at best, showed capitalism's inevitable collapse. Yet, the key problem with the theory of 'scientific communism' as a theory of planned socialized economy is that *there is none*.

As a result, the Soviets, as a form of allegedly proletarian self-governance, became a hostage to the dictatorial policies of the Bolshevik party. Martov, an early witness of the evolution of the Soviet system, points out that already in the early power struggle among the revolutionary factions, Lenin called to substitute the slogan "All Power to the Soviets" with "All Power to the Bolshevik Party." Only after the party squashed the opposition were the Soviets reinstated and lauded as a form of 'democratic' governance, assuming that political and economic power would firmly remain in the hands of the Bolsheviks.

The spirit of the pseudo-proletarian democracy (resonating with the attitude of some proponents of liberal democracy) is well expressed by the words of a well-known, at the time, Swiss socialist: "Considering that scientific socialism is truth itself, the minority understanding this truth is obligated to impose it on the masses";[50] in modern terminology, the 'power elites' are destined to lead the masses, while the appeal to the masses and critique of the elites is condemned as dangerous 'populism.'[51] In the methodological and political slippery slope of Marx's *humanism*, the orthodox (pseudo) Marxists re-delegated the role of the leading historical and social force, as well as the economic and social management functions, to the *revolutionary party* and state *bureaucracy*. As a result, as Martov observes, the emerging Soviet state was moving towards the growing centralization and development of the rigid bureaucratic apparatus, 'emancipated' from democratic control.[52] The strengthening new 'power elite' claimed to represent the proletariat in the final transition to communism and mold 'the last man' of the future as a new bearer of universal consciousness.

---

[50] Мартов, *Мировой Большевизм*, стр. 45. Quoted by Martov from Ch. Naine. "La dictature du proletariat et les ouvriers suisses."

[51] Let us recall that for Fukuyama a characteristic feature of populism is critique of the elites.

[52] Мартов, *Мировой Большевизм*, стр. 37.

Consequently, the 'liberated' workers became the clay in the hands of the self-proclaimed 'driver' of history and, in the name of the non-existent proletariat ('working class'), the party and state bureaucracy obtained the virtually dictatorial powers of the new secular priesthood. The state was de facto turned into the *church*, implementing in a grotesque form the system of what Mises calls *statism*. As history proved once and again, the idea of political leadership by the omniscient and omnipotent state or 'power elite,' no matter how noble and democratic the justification may sound, is a deadly prescription for totalitarian society, where, in a spirit of Hegel (and Plato), the State becomes a de facto definer and bearer of social freedoms and their indisputable enforcer.[53] In the Leninist-Stalinist practice, the ideal of historical emancipation of *human essence* gave way to the reality of the Orwellian totalitarian *existence*.[54] Methodologically, while we cannot enter the *theory* of humane society without Marx's essentialist concept of man as a universal being, we cannot retain this concept with the Marxist or pseudo-liberal state-focused socioeconomic *practice*.[55]

Kautsky and the other 'renegades' of the *Second International* foresaw the dangers of Lenin's "militant Marxism," eventually coming true in the Thermidor of the October Revolution and the consequent rise of Stalinism.[56] Arguing with Lenin, Kautsky insists that the 'dictatorship of the proletariat' should be understood as a rule of a democratic majority, possible only when capitalism reaches a high level of development (when the proletariat becomes a majority) and the transition to a new socioeconomic system can be accomplished peacefully, the idea Lenin fiercely opposed. A shroud tactician, Lenin saw (and contributed to) the political and social crisis triggered by the First World War, seeing it as an opportunity for a power grab. Theoretically, he justified his position by referring

---

[53] The related analysis is in Djilas, M. (1957). *New Class: An Analysis of the Communist System*. Pragier. For the orthodox exposition of the role of the state and the revolutionary party see Ленин, В. И. (1917). *Государство и Революция*; Lenin, V. I. (2011). *State and Revolution*. Martino Fine Books.

[54] In his insightful analysis of the twentieth century political landscape, Fromm explores the intrinsic link between human nature and historical dynamic: Fromm, E. (1941). *Escape from Freedom*.

[55] This problematic could be traced to Rousseau's idea of *general will*; in the *Confessions*, he remarks that the ultimate social objective is to find a societal form capable of *molding* the best and virtuous nation in the broad sense of the word; Fichte assumes an existence of such ('normal') nation *at the beginning* of historical time.

[56] Lenin uses the phrase "militant Marxism" in Lenin, V. I. (2103). *What Is to Be Done?* Martino Fine Books.

to the 'dialectical' character of the 'revolutionary situation' in Russia, when "the lower classes could not stand their impoverished condition any longer, and the upper classes could not hold to power any longer." The ensuing 'revolutionary situation' had to be resolved by the violent coup, in line with the 'dialectical law' of the transformation of quantity into quality.

In spite of their disagreements, the leaders of the *Second International* (including Kautsky) and Lenin joined ranks against the 'revisionism' of Edward Bernstein, a one-time close associate of Engels and an executor of his will. In *What to Be Done?* Lenin links Bernstein's ideas to the so-called 'trade unionism' and 'economism,' opposing both for disregarding the importance of militant Marxist social struggle. Bernstein (like Kautsky) argued against the Bolshevik-style revolutionary coup d'ètat and advocated liberal politics of the gradual amelioration of socioeconomic conditions, claiming that the reformist path was consistent with and reflected the true spirit of the Founders' views, especially in their late period.[57] While agreeing with his opponents that a peaceful transition to socialism, in particular in the United States and Britain, was something that Marx and Engels considered as an option in the 1870th, Lenin claimed that this path was no longer viable in the early twentieth century because capitalism entered into its imperialist, militant phase and could be overthrown only by force. It is germane to note that, even apart from their political evolution, Marx and Engels, as early as 1846, defined communism as "not a state of affairs which is to be established, but an ideal to which reality [will] have to adjust itself." In the spirit of Fichte, they emphasized the historical process itself rather than the allegedly final outcome of history, explicitly stating: "We call communism the real movement which abolishes the present state of things."[58]

On the heels of Lenin's political success, *Marxism-Leninism* became a dominant representative of the Marxist orthodoxy that allegedly followed and preserved the spirit and the letter of Marx's teaching. Using dialectic as a tool of political struggle, Lenin pronounces it a universal conceptual instrument of Marxism, emphasizing that one cannot fully understand

---

[57] Bernstein, E. (1899). *Voraussetzungen des Sozialismus and die Aufgaben der Sozialdemokratie.* Translated as Bernstein, E. (1993). *The Preconditions for Socialism.* Cambridge University Press. See Lenin's critique of Bernstein in Ленин, В. И. (1903). *Что Делать?* (Lenin, V. I. *What to be Done?*).

[58] Marx, K. and Engels, F. (1998). *The German Ideology,* p. 57.

Marx's *Capital* without a thorough knowledge of Hegel's *Science of Logic*. The allegedly materialist dialectical method becomes an essential ingredient of, and the methodological link between, the orthodox conceptions of *dialectical* and *historical* materialism.

Towards Dialectical Materialism

While Marx's teaching represents an allegedly integral political, social, and philosophical doctrine, it is divided, by its orthodox adherents, into *historical materialism*, covering issues related to the social and economic transition to communism, and *dialectical materialism*, considering traditional philosophical problems, such as a relationship between matter and spirit, freedom and necessity, or the nature of truth. At the core of dialectical materialism is an alleged synthesis of the principles of *dialectics* with the notion of *matter*. In this respect, an important role in the codification of Marxist orthodoxy belongs to G. V. Plekhanov (1857–1918), one of the 'founding fathers' of Russian Marxism, who claimed that in philosophy Marx and Engels adopted a standpoint of Spinozism freed from its theological disguise by Feuerbach.[59] To the extent that for Spinoza God is not a *transiens* cause of nature but *immanens* one, the two represent a totality of *substance* that could be viewed as a cause of itself[60]; the orthodox Marxists view this totality as *matter*. For dialectical materialism, the substance is in a constant state of self-motion, producing, according to the dialectical laws, a variety of material phenomena. It is from this perspective that the pantheism of Spinoza is (re)interpreted by Plekhanov as new materialism. Notwithstanding Plekhanov's opposition to Bolshevism, Lenin supported his version of materialism. In the spirit of new Spinozism,

---

[59] Plekhanov, G. V. (1992). *Fundamental Problems of Marxism* (2nd ed.). Intl Pub Co Inc. As a politician, Plekhanov joined the Menshevik (minority) opposition, consequently crashed by Lenin and his comrades. About Plekhanov see Baron, S. H. (1963). *Plekhanov. The Father of Russian Marxism*. Stanford. A brief discussion of his views is in Kolakowski, L. (1978). *Main Currents of Marxism* (Vol. 2) (S. Falla. Trans.). Clarendon Press; see also Zen'kovski, V. V. (1989). *Istoriya Russkoy Filosofiyi* (Vol. 2) (2nd ed.). Paris: YMCA-Press; translated as Zenkovsky, V. V. (1953). *A History of Russian Philosophy* (Vol. 1–2) (G. L. Kline. Trans.). Routledge and Kegan Paul. On Lenin see, for example Shub, D. (1966). *Lenin. A Biography*. London.

[60] Spinoza, B. *Ethics*, part I, theorem 18.

he insists that "it is necessary to deepen the knowledge of matter to the knowledge (notion) of substance."[61]

Such interpretation of Marx's philosophy could be traced to Engels' attempt to integrate 'contemplative' materialism and dialectics in the unfinished *Dialectics of Nature*.[62] In the Hegelian-style 'natural philosophy,' he imposes a speculative dialectical schema on reality and introduces the three dialectical laws: the transformation of quantity into quality, the mutual penetration of the opposites, and the negation of the negation. In its dialectical 'self-motion,' 'the continuously changing world of material forms' appears to us as *objective reality*, a foundational notion of the new materialism.

The objectivity of the material world, one of the central issues in Lenin's *Materialism and Empiriocriticism* (1909), has been a recurrent theme in the debates over Marx's philosophy. For example, A. Schmidt maintains that already in the 1844 *Manuscripts*, Marx "adopted... an objective viewpoint," "similar" to Lenin's view on matter as 'objective reality.'[63] To support this interpretation, he quotes Marx: "A being which does not have its nature outside itself is not a **natural** being and does not share in the being of nature." In Schmidt's reading, man is viewed as a natural being to the extent that he *belongs* to nature as it exists 'in itself,' even though the latter expression is not used. As we could see earlier, such a position is inconsistent with Marx's anthropocentric ontology and his conception of man as a universal being (and, consequently, a universal subject of history). Marx's reference to a being having a "subjective nature outside itself" points to a coextensive relationship between man and his *extended body* of nature, a view more in line with Kantian transcendentalism than traditional metaphysics in general and Spinozism in particular.

Trying to resolve the conundrum of anthropological nature, another researcher interprets it as a conceptual *deficiency* of Marx's idea of

---

[61] Lenin (1963). *Filosofskie Tetradi (Philosophical Notebooks)* In Lenin, V. I. *Polnoye Sobranie Sochineniy (Collected Works)* (Vol. 29, p.142). Moskva: Izdatelstvo. Politicheskoy. Literatury. Plekhanov, G. V. (1906). *Kritika Nashich Kritikov (Critique of Our Critics)* (p. 166). St. Petersburg: Obshchestvennaya Polza.

[62] Engels, F. (1978). *Dialectics of Nature* In Marx, & K Engels, F. *Collected Works* (Vol. 25). International Pub. Co. Marx's dialectical 'anthropocentrism' should not be identified with Engels' dialectic of nature, which represents a little more than an old-fashioned *Naturphilosophie* in the materialist disguise. For a brief discussion, see Kolakowski, L. (1978). *Main Currents in Marxism*. (Vol. 1). Oxford University Press.

[63] Schmidt, A. (1971). *The Concept of Nature in Marx* (p. 64) (Fowkes. Trans.). B. NBL.

ontological unity between subjective and objective nature: "There is something unsatisfactory about talk of a 'complete unity in essence of man with nature.' There is indeed something deficient in the ontology of the 1844 *Manuscripts*. In my view, the problem has to do less with the assimilation of Hegel than with that of Feuerbach."[64] Such comments reflect only an incomplete understanding of Marx and misconstrue the very spirit of his social anthropology. By contrast with dialectical materialism, Marx maintains that man has his (subjective) nature outside himself *to the extent that it emerges for him* through his consequent instrumental relation to (objective) nature. To reiterate, for Marx, it is the coextensive relationship between man and "the whole of nature" that constitutes the *objectivity (universality)* of man as a total or truly human being. Such an understanding of objectivity goes back to Kant, who explains that "pure reason requires completeness in the application of the understanding in the system of experience," and "this completeness is presented by reason as cognition of objects." He concludes that the "object is only an idea intended to bring the cognitive act of the understanding as close as possible to the completeness expressed by the idea."[65] Thus, the term *object* refers not to the material source of human experience in itself ('substance') but to the formal (aka objective) way sensual data are presented *for man* in the course of his (cognitive) activity.

In his *inversion of Kant*, Marx construes the *objectivity* of human experience not in the context of formal requirements of pure reason but through the instrumental activity of collective labor that constitutes nature as a human extended body and man as a *being for himself* and, thus, an *objective being*. In this context, Marx's concept of objectivity refers to the *completeness of human experience* as it emerges in the collective instrumental activity; in a different terminological framework, the objectivity of human experience means that man and nature belong to the same class (category) closed under the *invariant operation* of the instrumental activity. In this sense, ontology (of nature) recapitulates human technological practice. Needless to say, this view pointedly discounts the orthodox metaphysics of nature as objective (in itself) reality, substance, etc.

In the final analysis, dialectical materialism is incompatible with Marx's philosophy of praxis and appears as a form of contemplative (and hence abstract, inconsistent) materialism referred to by Marx in his critique of

---

[64] Arthur, C. J. (1986). *Dialectics of Labor*, p. 133.
[65] Kant, I. *Prolegomena to Any Future Metaphysics*, part 3, par. 44.

Feuerbach. The concept of *substance* represents nothing but a secular symbol of the faith of the Hegelian-style esoteric theology. In fact, Lenin's call to 'deepen' the notion of matter to the notion of substance is at odds even with the strong, though not always consistent, anti-metaphysical trend in Engels's thought. In the *Dialectics of Nature*, he says: "The materialistic outlook on nature means nothing more than the simple conception of nature just as it is, without alien addition." Engels clarifies further that this "simple conception of nature" is based on the notion of "reciprocal action": "We cannot back further than to knowledge of this reciprocal action, for the very reason that there is nothing behind to know" (beyond what is revealed for man in his instrumental interaction with nature).[66] It is the anthropocentric focus of Marx's social ontology that finds development in the allegedly 'revisionist' interpretations of his humanism.

PHILOSOPHY OF PRAXIS VS. DIALECTICAL MATERIALISM

The strongest conceptual support for the non-orthodox interpretations of Marx could be found in his early writings, especially the *Economic and Philosophical Manuscripts*. However, the 1844 *Paris Manuscripts* and the *German Ideology* became available only in 1932, when dialectical and historical materialism had already been codified as genuine forms of Marxism. To a large extent, it was Kant who became an inspiration for those who wanted to part the company with orthodox Marxism, particularly in the citadel of Marxist orthodoxy. As observed by Zenkovsky, "Russian followers of Marx, who later became extremely productive in philosophy (Bulgakov, Frank, Berdyaev and Struve), from the very beginning aimed at the synthesis of historical materialism with Kantianism."[67] However, after the 1917 revolution, they completely broke with Marxism, and Kantian criticism retained its influence on Marx's followers mainly in Germany and Austria. For example, a noted Austro-Marxist Max Adler (1873–1937) links Marx's "concept of socialized man" to "transcendentally socialized consciousness":

> It is Marxism which makes possible a more consistent development of the epoch-making trend of German philosophy, in which the unique character

---

[66] Engels, F. (1954). *Dialectics of Nature*. (p. 307). Moscow: Foreign Languages Publishing House.
[67] Zenkovsky. (1953), vol. 2, p. 273.

of social consciousness was elaborated. The means to do that is the basic concept of **socialized** man which Marx developed.... The historical socialization of this socialized man is only possible where the individual consciousness is already transcendentally socialized. Social association... is a transcendental condition of experience. Thus, social experience is grounded as experience of **being** in exactly the same way as experience of nature.[68]

Adler (rightly) places Marx in the tradition of German critical philosophy and claims that his achievement was to demonstrate a unity of transcendental experience in its social and natural modes. Likewise, Jurgen Habermas refers to the Kantian conception of transcendental apperception as a precursor of Marx's understanding of experience: "Like Kant's original apperception, the materialist conception of synthesis preserves the difference between form and matter. Of course, the forms are categories not primarily of the understanding but of objective activity; and the unity of objectivity of possible objects of experience is formed not in transcendental consciousness but in the behavioral system of instrumental action."[69] In this interpretation, the epistemic synthesis of 'pure' reason is replaced by the activity of productive labor which, in conformity with Kantian criticism, constitutes the "objectivity of... experience."

On this background, it is understandable that Lenin, in *Materialism and Empiriocriticism* (1909), targets a wide range of what he views as idealist distortions of genuine materialism and traces them to Kantian transcendentalism. In his view, Feuerbach, Marx, and Engels criticized Kant "from the left," while 'revisionists' criticized him "from the right," returning to the empiricism of Hume and the subjective idealism of Berkeley. In this work, the principal target of Lenin's critique became the so-called *empiriocriticism*, represented most prominently by Ernst Mach (1838–1916) and Richard Avenarius (1843–1896): "The whole *school* of Feuerbach, Marx and Engels moved to the left of Kant, towards the complete rejection of any idealism and agnosticism, while our Machists went into the different direction of the *reactionary* movement in philosophy,

---

[68] Adler, M. (1978). *A Critique of Othmar Spann's Sociology* (*Zur Kritik der Soziologie Othmar Spann*) (1927) (p. 75) In *Austro-Marxism* (T. Bottomore and P. Goode. Trans.). Clarendon Press.

[69] Habermas, J. (1971). *Knowledge and Human Interests*, p. 34.

following Mach and Avenarius who criticized Kant from the standpoint of Hume and Berkeley."[70]

To a large extent, the fierce debate over seemingly abstract issues of metaphysics was inspired by the early twentieth-century scientific revolution, including Einstein's relativity theory and the emerging quantum physics. This discussion allows Lenin to reiterate his views on the notions of matter, objective reality, truth, etc. Apart from attacking (on the verge of verbal abuse) the founders of the movement, his criticism is directed at the highly original interpretation of Marx by A. Bogdanov (Malinovsky) (1873–1928), his political ally and one of the leading Russian Marxists of the early twentieth century.[71] Zenkovsky (a religious thinker, far away from Marxism) pays tribute to Bogdanov as one of the most original philosophers in the history of Marxism: "Bogdanov always remains a **free** thinker, having accepted Marxism sincerely and seriously; that is why he is inclined to the most radical revisionism." Bogdanov's philosophy is especially interesting as being perhaps the most consistent (and, indeed, radical) interpretation of Marx's anthropocentric methodology. Though his approach echoes the spirit of Marx's early works, neither the 1844 *Manuscripts* nor *German Ideology* was available to Bogdanov, who died in 1928.

Conceptually, Bogdanov draws together the pragmatic reading of Marx by Labriola and Brzozowski, the positivism of Spencer, and the empiriocriticism of Mach. In the framework of empiriocriticism, reality is viewed as a combination of neutral 'elements of experience.'[72] *Relative to the position of an observer*, certain 'complexes' can be qualified as either physical or psychical, objective or subjective. On this background, Bogdanov develops

---

[70] Lenin. *Materialism and Empiriocriticism* In: Lenin, V. I. *Collected Works* (Vol. 18, Chap. 5). For the opposing view, see Mach, E. (1906). *Die Analyse der Empfindungen*. Jena. Among Mach's notable ideas is a *relativity principle* that contributed to the development of Einstein's general relativity theory.

[71] Zenkovsky, V. V. (1953), vol. 2, p. 281. Zenkowsky gives a good but sketchy outline of Bogdanov's views. Kolakowski discusses mostly only one of Bogdanov's work—*Empiriomonism*. Yassour (1983) restates the views of Bogdanov, Lenin, and Plekhanov without in-depth analysis of the philosophical controversy. See also Kats, Y. (2004). "Bogdanov, Marx and the Limits to Growth Debate." *European Legacy*, 9 (3), pp. 301–317. A trained physician, he was a director of the Moscow Institute for the Transfusion of Blood and died after performing an experiment on himself. His major works listed in the bibliography.

[72] See Mach, E. (1906). *Die Analyse der Empfindungen*. Jena. Mach's concept of the 'elements of experience' conspicuously influenced some members of Vienna Circle, as well as Russell, in their treatment of 'sense-data.'

his version of empiriocriticism—*empiriomonism*—and argues that "the system of experience is the system of labor; all its content remains in a framework of collective human activity."[73] As an abstraction of substance is eliminated through the idea of human praxis, *nature* is understood not through the 'contemplative' notion of matter but through the instrumental activity of the "infinitely unfolding domain of man's labor."[74] Paraphrasing Mill, we can say that for Bogdanov, nature is a permanent possibility of technology-based (collective) labor that determines, in the final analysis, 'what there is.' In a sense, *ontology recapitulates technology*. Curiously, the idea of the elements of experience and their 'complexes' constituting the 'objectivity' of human experience resonates with the Buddhist notion of 'dharma,' though Buddhism understands it in a purely psychological context. If Marx's and Bogdanov's ontology recapitulates technological practice, Buddhist *ontology recapitulates psychology*.

Bogdanov attacks contemplative materialism in general, and the idea of substance in particular, by utilizing the notion of *fetishism*: "Scientific philosophy takes the very notion of an 'absolute' as one of the **cognitive fetishes**." He explains that scientific fetishism is similar to the fetishism of commodities, where social relations between people are perceived as relations between commodities. For example, due to the fetishism of *matter*, "the relation between activity and matter becomes completely perverted: activity turns out to be an intrinsic property of matter [and] a notion of labor is made dependent on the notion of matter, while in reality matter depends on labor."[75] In the spirit of Fichte, 'matter' appears as a permanent possibility of historically unfolding labor activities: "the idea of 'matter' is in correspondence with **collective** labor... Generally speaking, matter is an **intersubjective** (obscheznachimoye) resistance to human efforts, the object of **collective** labor."[76]

Developing this argument, Bogdanov explains that the origin of *cognitive fetishism* lies in the alienation of an individual from society. An individual falsely perceives 'matter' as an objective reality to the extent that, in the capitalist society, he is alienated from the collective labor practices (and consequently, from his true social essence). Accordingly, *social*

---

[73] Bogdanov. (1913). *Philosophy of Vital Experience*, p. 83.
[74] Bogdanov. (1913), p. 56.
[75] Bogdanov. (1913). *Philosophy of Vital Experience*, pp. 73–74. On fetishism of commodities see our discussion of Hess and Marx in the previous chapter.
[76] Bogdanov. (1913), p. 58; in Brzozowski's words, nature is "coextensive with labor"; see Kolakowski, vol. 2, p. 226.

emancipation would entail *epistemological emancipation*: as soon as an individual identifies his labor with collective labor, he sees 'matter' as a correlative 'resistance' to labor practices rather than a mysterious material substance 'in itself.' Needless to say, this view directly contradicts the philosophy of dialectical materialism.

Bogdanov further explicates the concept of cognitive fetishism, developing a unique synthesis of Mach's empiricism and Marx's philosophy of praxis. He sees the core of Mach's methodology in exposing the *fetishism of physical notions*, which convey an idea of substance in one form or another. As an example, Bogdanov considers geometry, which "serves for a collective as a tool organizing, in a certain respect, its practice."[77] In line with Einstein's general relativity theory, he maintains that geometrical theorems 'in themselves' are neither true nor false: their truth value depends on the procedure of measurement carried out in the gravitational field. Thus, concludes Bogdanov, the concept of the Euclidean absolute space is nothing but a fetish; another example of a cognitive fetish is a purely relational notion of mass which, in accordance with *Mach's principle*, "does not express anything but a resistance to action."[78] Bogdanov's approach is also in tune with the emerging, at the dawn of the twentieth-century, quantum physics. The quantum-mechanical picture of reality (particularly Heisenberg's uncertainty principle) is hard to reconcile with Lenin's naïve 'theory of reflection.' By contrast, new physics perfectly fits Bogdanov's (and Marx's) picture of the world: any 'objective' description of natural phenomena is possible only in the context of human instrumental activity.

Such an approach is exactly what Marx's revolution in philosophy means for Bogdanov: reality can be understood (and changed) only by viewing human activity as a constitutive element of the world. Extending this approach to epistemology, Bogdanov claims that the concepts of 'truth in itself' and 'pure theory' are meaningless. Knowledge is *pragmatic*, though not entirely conventional: the objectivity of knowledge arises from the collective or 'inter-subjective' nature of praxis.

For both the orthodox and revisionist brands of Marxism, the ideologeme of infinite growth is linked to the corresponding epistemological presuppositions. In the framework of Lenin's orthodoxy, infinite (scientific and technological) growth is conceptually secured by the

[77] Bogdanov. (1913). *Philosophy of Vital Experience*, p. 81.
[78] Bogdanov. (1913). *Philosophy of Vital Experience*, p. 79.

assumingly "infinite process of deepening of human cognition of things, appearances, processes etc., from appearance to essence and from less deep essence to deeper essence."[79] As we discussed, such a claim is inconsistent with Marx's anthropocentric ontology of labor. Moreover, the infinite and ever-deepening cognition of the allegedly inexhaustible *material substance* would be possible only for the infinite *thinking substance*, thus undermining the consistency of dialectical materialism and turning it into a version of either Cartesian dualism or a straightforward Spinozism (under a dialectical guise). In socioeconomic terms, the old-fashioned ontology of infinite substance is translated into an assumption of an unlimited resource base and a social *telos* of ever-expanding material production.

By contrast, Bogdanov construes the 'ever-increasing reign' over nature in the framework of his unique project of unified science—the *general theory of organization* called Tektology (from Greek 'tekto'—organization). The latter follows the general presuppositions of his philosophy, combining the elements of empiriocriticism, Spencer's universal evolution, and Marx's philosophy of praxis. We start the next chapter with a discussion of Bogdanov's project of unified science and examine his proof of 'inexhaustible creativity' based on the 'tektological' interpretation of the second law of thermodynamics. Bogdanov's proof leads us to the analysis of the 'limits to growth' and 'end of science' debates, where we proceed beyond Marxism and explore contemporary arguments for and against a postulate of infinite scientific and technological growth.

[79] Lenin. (1963). *Philosophical Notebooks*, p. 203.

CHAPTER 5

Towards the Critique of Technological Reason

This creating, willing, valuing ego, which is the measure and value of all things
—Nietzsche

The beginning of the twentieth century, a time of global economic instability and political unrest, culminated in the First World War and the 1917 Russian Revolution. While the accelerating social discontent and ideological strife dealt a blow to the historical optimism of the Enlightenment, the rapid scientific advancements undermined the old-standing scientific paradigm, inspiring a quest for the new conceptual foundations of natural sciences and mathematics. Questions like the 'disappearance' of matter, the elusive objectivity of space and time, and new forms of natural causality came to the forefront of philosophical discussions. Elaborating in his 1922 letter on the philosophical significance of relativistic physics and quantum theory, Herman Weyl declares that physical "laws do not matter for the essential contents of reality; the ground of reality is not grasped by them." On the same occasion, Einstein expresses his appreciation of Poincare's conventionalism, stating that a conceptual framework is a product of "free choice from the point of view of logical completeness and adaptability to experience, as indeed is so beautifully shown [by] Henri Poincarè."[1] The split between straightforward rationalism and the ideal of science advanced by the Enlightenment vs. the spirit of the arriving epoch is conspicuously reflected in the 1929 Davos debate between Cassirer and Heidegger. At stake is a dichotomy of the rationally organized *objective reality* vs. the existential openness of human being-in-the-world, what Heidegger calls *Dasein*, as a principal constitutive factor of Being.

With relativism in science, philosophy, and art ruling the day, *empiriocriticism*, claiming to have reconciled traditional metaphysics with a newly evolving scientific paradigm, becomes an especially popular conceptual choice for both working scientists and scientifically minded philosophers. In this vein, A. Bogdanov, a revolutionary in politics and science, applies the principles of empiriocriticism to Marx's philosophy of praxis and redefines 'materialistic' dialectic in the framework of yet another *Scientia Universalis*—a *general theory of organization* (*Tektology*), considered by some as a precursor of *cybernetics* and the *general systems theory*. Using his 'tektological' interpretation of the second law of thermodynamics, Bogdanov develops a case for the *inexhaustible creativity* of mankind as an engine of an open-ended historical advancement to the social 'realm of freedom.' However, the eclectic synthesis of Mach and Marx does not allow him to maintain a consistently scientific standpoint, and the proof of infinite creativity exposes its Marxist ideological underbelly. At the same time, the innovative character of Bogdanov's methodology resonates with a pragmatic focus and the scientific spirit of the twentieth-century debates over the limits to technological and scientific progress.

After the Second World War, the renewed expectations of continuous economic growth and the emergence of a globalized economy encouraged the Neo-Malthusians to voice new concerns regarding the side effects of uncontrolled technological and economic expansion, such as environmental damage, overpopulation, and potential depletion of world resource-base. For example, a group of researchers associated with the Club of Rome, utilizing a comprehensive *systems approach*, argued back in

---

[1] Citations from Weyl and Einstein are from Weyl, H. (2009). *Mind and Nature: Selected Writings on Philosophy, Mathematics and Physics* (pp. 26–27) (P. Pesic. Ed.). Princeton University Press. Poincare, H. (2018). *The Value of Science* (ch. 7). Forgotten Books (Originally published 1900). For a Neo-Kantian view on science, see Cassirer, E. (1923). *Function and Substance and Einstein Theory of Relativity* (W. C. Swaby and M. C. Swaby. Trans.). The Open Court Pub. The *Function and Substance* became a precursor of his *philosophy of symbolic forms*.

the 1970th that, by the year 2100, the world would have reached the limits of economic expansion, entering into a state of deepening downturn, accompanied by a drastic decline in world population and industrial capacity.[2] On this background, the authors emphasize the urgency of adopting the far-reaching 'zero growth' policies to be implemented on a global scale. While this potential 'gloom and doom' scenario is supported by sound statistical analysis and advanced computer modeling techniques, the authors do not contest, *as a matter* of *principle,* the unlimited potential of technological progress. The related attempts to consider *theoretical limits* to technological growth do not go much beyond the references to the second law of thermodynamics, already used by Bogdanov in the 1920th (though with a directly opposite conclusion). By keeping the belief in *theoretically* unlimited technological progress virtually unchallenged, the authors leave a conceptual gap in their model, undermining the predictive power of zero growth policy recommendations.

The typical arguments for and against infinite *scientific* progress range from practical to strictly theoretical. Among the practical causes for a potential 'saturation' of science are diminishing intellectual returns, cultural changes, and economic problems. The theoretical arguments utilize a wide range of principles, from the familiar second law of thermodynamics to fundamental results in (meta) mathematics, such as Gödel's incompleteness theorem. Methodologically, such arguments appear incomplete or inconsistent, recapitulating the Cartesian dualism of knowledge and being. Historically, they are rooted in the ideological premises of the Age of Reason, with its secularized religion of infinite growth and the unshakable belief in the universality of man as a master and possessor of nature. Ultimately, the problematic of limits to scientific and technological progress calls for a theoretical investigation of its epistemic foundations. In Kant's terminology, it is an issue of a systematic critique of *instrumental reason*, its scope and extent, practical applicability, and value structure.

---

[2] Meadows, D. H. et all. (1972). *The Limits to Growth.* Universe Books. The "system dynamics" methodology is based on the works of MIT Prof Jay Forrester. The authors of the *Limits to Growth* call him the "intellectual farther" of their model. Forrester, J. W. (1971). *World Dynamics.* Wright-Allen Press.

From Philosophy of Praxis to General Theory of Organization

Bogdanov's project of unified science—*Tektology* (from the Greek *tekto*—organization)—is an attempt to develop a new 'dialectic of nature' "embracing both practical and theoretical methods, purposeful human methods as well as spontaneous methods of nature."[3] In line with the principles of Marx's philosophy of praxis, Bogdanov maintains that "the system of experience... remains in a framework of collective human activity" and adds that "human activity—from the most elementary to the most complex forms—is reducible to the processes of organization";[4] consequently, "all the interests of mankind are organizational. Hence, there cannot and must not be any other worldview than organizational."[5] In a brief historical account of *universal science*, Bogdanov mentions Hegel, Marx, and Spencer, claiming that "the first attempt at **universal** methodology belongs to Hegel. In his **dialectic,** he thought to find a general universal [mirovoi] method understood... as a method of 'development.'" Though Hegel's approach was too "unclear and abstract," continues Bogdanov, it made "a strong impact on the further progress of the organizational ideas. Spencer's system of universal evolution, and especially materialist dialectic, became the next approximations to the present [tektological] approach."[6]

While some present *Tektology* as the first version of the *general system theory* of Ludwig von Bertalanffy or Norbert Wiener's *cybernetics*, it was Herbert Spencer who articulated the key principles of a modern *theory of automatic control*, including the notions of system, organization, negative and positive feedback, dynamic equilibrium, etc.[7] While Spencer's approach is often interpreted as a crude reductionism of social phenomena to biological ones, nothing is further from the truth. As a matter of fact, Spencer explicitly criticizes what he considers as the reductionism of Plato and Hobbes: the first one—for the attempt to draw a straightforward analogy between society and the human mind, the second—for the attempt to

---

[3] Bogdanov, A. (1989). *Tektologiya: Vseobschaya Organizatzionnaya Nauka* [Tektology: The General Science of Organization] (Vol. 1, p. 112). Moskva: Ekonomika.

[4] Bogdanov *Philosophy of Vital Experience*, p. 83.

[5] Bogdanov, A. (1989). *Tektologiya*, vol. 1, pp. 70–71.

[6] Bogdanov, A. (1989). *Tektologiya*, vol. 1, p. 112.

[7] Bertalanffy, L. (1969). *General System Theory*. George Braziller Inc. Winer, N. (2013). *Cybernetics*. (2nd ed) Martino Fine Books.

draw a direct analogy between society and the human body. Spencer argues that the "chief errors of these comparisons made by Plato and Hobbes... [is that] both thinkers assume that the <u>organization of a society is comparable</u>, <u>not simply to the organization</u> of a living body <u>in general</u>, but to the organization of the human body <u>in particular</u>."[8] He emphasizes that "lacking the great generalizations of biology, it was... impossible to trace out the real relations of social organizations to organizations of <u>another order</u>" and assesses any direct analogy between particular biological and social entities as a form of (invalid) reductionism.[9]

In line with the systems approach, Spencer's conception of universal evolution is construed as a process involving the "continually increasing complexity of structure," when "parts gradually acquire a mutual dependence," characterized as the development and "maintenance of a controlling apparatus." In both social and biological organizations, the crucial characteristic of such an apparatus is the **"positively regulative** and the **negatively regulative**" systems of control or "those which stimulate and direct, as distinguished from those which simply restrain."[10] Spencer makes clear that he is talking about "organizations of every kind" and emphasizes the importance of central for the functionality of any stable dynamic system principle of "negatively regulative control";[11] needless to say, negative feedback is a fundamental principle of modern automatic control theory (cybernetics). Spencer productively utilized principles of automatic control in applied research, including analysis of biological populations and social phenomena.[12]

While Spencer developed his theory of universal evolution in a predominantly scientific framework, Bogdanov's project of universal science is heavily influenced by Marx's anthropocentric methodology 'enriched' by the empiriocriticism of Mach. To a large extent, the principles of *Tektology* are based on the dialectical laws interpreted from the perspective of the allegedly universal organizational methods. The 'penetration of the

---

[8] Spencer, H. (1981). "The Social Organism" (1860) In Spencer, H. *The Man Versus the State*. (pp. 390–391). Liberty Classics.

[9] Spencer (1981). pp. 391–392.

[10] Spencer, H. (1981). "Specialized Administration" In Spencer, H. *The Man Versus the State*, p.456.

[11] Spencer, H. "Specialized Administration," pp. 481–482.

[12] For an assessment of Spencer's contribution to theoretical biology see: Zavadsky K. M. and Kolchinsky E. I. (1977). *Evolutziya Evolutzii*. [Evolution of the Evolution] (pp. 167–171) Leningrad: Nauka.

opposites' is substituted by the dynamic equilibrium of the opposing material forces; the 'transformation of quantity into quality'—by a 'crisis' signifying the transformation of one organizational form into another (for example, freezing water changes into ice); and the 'negation of the negation' refers to the correspondence between the 'lower' and 'higher' levels of dynamic equilibrium.

The 'negation of the negation' law has always been controversial in Marxist literature. Within the Hegelian system, it entails a close-ended dialectic of history, while in its Marxian inversion (as we argued)—a close-ended dialectic of labor and technology. By contrast, for Bogdanov, the state of 'synthesis of the opposites' represents only an intermediate stage of dynamic equilibrium constituted by the oscillation of the opposing forces, no matter what their nature may be; in a way, this process is similar to the continuous transition of the well-designed automatic control system from one stable state to another. From a different (but related) perspective, Bogdanov's interpretation of dialectical 'negation' as a dynamic equilibrium resonates with Spencer's 'law of rhythm' (productively applied by Spencer in the analysis of biological populations). In this framework, the 'dialectic of history' remains an open-ended process, as there is nothing to be revealed from any formally defined *in itself* state of affairs to its *for itself* externalized form. In an attempt to demonstrate the open-ended nature of human history by utilizing the principles of his general theory of organization, Bogdanov argues that the driving force of the historical process is the 'inexhaustible creativity' of mankind.

INEXHAUSTIBLE CREATIVITY AND INFINITE GROWTH

Bogdanov's argument for the "*ever-increasing reign of mankind over nature* by means of machine industry" follows the basic presupposition of Marx's philosophy of history.[13] Based on the *second law of thermodynamics* and the so-called *selection principle*, the argument for 'inexhaustible creativity' is historically and conceptually significant due to the apparent shift from an abstract metaphysics of traditional philosophical discourse to the conceptual framework of modern science.[14]

---

[13] Bogdanov, A. (1913). *Filosofiya z'ivogo opyta* [Philosophy of Vital Experience], p. 274. See the last section of the previous chapter.

[14] The application of the selection principle in the context of thermodynamics resonates with some ideas of Ilya Prigogine and his school. Prigogine, I. and George, C. (1983). "The

According to the second law, in a *closed system*, there is a continuous and irreversible loss of usable energy, with a corresponding increase of *entropy* measure. Eventually, when the entropy reaches its maximum, the system enters the 'death' state of thermodynamic equilibrium. If we accept that scientific and technological practice (and human activity in general) is inseparable from the different forms of the "interchange with Nature" (consumption of energy and raw materials, pollution, etc.), such activity is subject to the limitations of the second law as any other physical process.[15] From this perspective, technological progress would inevitably reach a state of (metaphorically speaking) 'technological equilibrium' when the 'entropy' of human civilization is maximized. In practice, this might be exhibited by insurmountable environmental problems and the inability to overcome them through further technological advancements.

Bogdanov's proof of the *inexhaustible* human creativity and mastery over nature combines the second law of thermodynamics and the 'selection principle.' Following Spencer, he maintains that the selection principle universally applies to mechanical motion, biological evolution and technological development. With an implicit reference to Mach, he further claims that any process can be viewed as a *positive* or a *negative* selection of the *elements of experience*. In this context, the negative selection, understood as a disintegration of (material) 'complexes,' is linked to the loss of usable energy (and the corresponding increase of entropy).[16] However, Bogdanov argues that the unavoidable loss of energy through negative selection could be replenished through 'progressive' (positive) selection, leading to the creation of more and more organized complexes (of experience). Among the prime examples of positive selection are biological processes and human creative (including technological) practices represented by, using Marx's expression, the "infinitely unfolding domain of man's labor."[17] The crucial point of his argument is that the *irreversible* character of physical processes is compensated by the *inexhaustible recreation* of nature (by human labor).

Second Law as a Selection Principle: The Microscopic Theory of Dissipative Processes in Quantum Systems." *Proceedings of the National Academy of Sciences.* Vol. 80, pp. 4590–4594. An attempt to apply these ideas to biology and social sciences is presented in Prigogine, I. and Stengers, I. (1984). *Order out of Chaos.* Heinemann.

[15] Marx, K. (1967). *Capital* (Vol. 3, p. 820) (F. Engels. Ed.). International Publishers,
[16] Bogdanov, A. (1989) *Tektologiya* [Tektology] (Vol. 2, p. 204).
[17] For related ideas see Marx, K. (1973). *Grundrisse* (pp. 56, 704) (M. Nicolaus. Trans.). Vintage Books. Marx, K. (1967). *Capital.*,vol. 3, p. 820.

The main problem with this line of argument is whether the progressive 'positive selection' occurs in an open or closed system. The second law applies to any physical, biological, or technical, *closed system*, where the measure of entropy would always increase, with an ultimate halt to any sort of 'creativity.' Bogdanov's argument could be salvaged only under an implicit assumption that *cognitive creativity* (as a form of progressive selection) construes the human environment as an *open system*. Then, in such an *epistemologically open system*, cognitive/technological practices would indeed not fall under the scope of the second law of thermodynamics and could proceed indefinitely.

However, such an (epistemic) line of reasoning would undermine the scientific character of Bogdanov's argument. If *physical limitations do not necessarily entail epistemological limitations*, infinite technological progress is not a matter of natural science but an *epistemic* question logically independent of any general physical principle. From this perspective, Bogdanov's argument would become circular because, by introducing the premise of epistemic infinity, the argument essentially circumvents the second law used to prove the 'inexhaustible creativity' in the first place. In such a logical framework, infinite creativity appears as an *epistemologically independent axiom*, making the second law at best redundant. Moreover, judging by his sparse comments, Bogdanov may be confusing the irreversible *nature of time* with an allegedly inexhaustible nature of *human creativity*. The irreversible character of thermodynamic processes has been indeed used to explain the direction of time from the past to the future. However, the analogy between time and human creativity is entirely unjustifiable in this context.[18]

In the final analysis, Bogdanov's contention of infinite growth remains an article of faith rather than an outcome of rigorous scientific analysis. His argument is inconsistent due to his attempt to combine a strictly scientific methodology with Marx's 'progressive' historical outlook. The thesis of 'inexhaustible creativity' is an implicit *premise* of the proof of infinite 'positive selection' (whatever form it may take) and relies on ad hoc epistemic assumptions independent of any physical theory.[19] The tension

---

[18] For the thermodynamic approach to time see Prigogine, I. Stengers, E. et al. (1984). *Order out of Chaos*. Bantam. Reichenbach, H. (1999). *The Direction of Time*. Dover Pub. (Originally published 1958).

[19] Bogdanov, A. (1989). *Tektologiya*, vol. 2, p. 207. His attempt to mix ideology with science also leads to mistakes in the interpretation of Einstein's theory of relativity.

between the epistemic and scientific aspects of Bogdanov's approach may be indicative of the general problem: any attempt to prove the (value-based) thesis of infinite scientific and technological progress (implied by inexhaustible creativity) within the limits of natural science would be either incomplete or inconsistent. What is required is a meta-scientific approach independent of physical theory in general and thermodynamics in particular.

Technology and the Limits to Growth

The idea that mankind may reach a critical point, when overpopulation and shortages of food, raw materials and energy resources would endanger the very existence of civilization, is not new and goes back at least to Thomas Malthus (1766–1834), who maintained that "population, when unchecked increases in a geometrical ratio. Subsistence increases only in an arithmetical ratio."[20] The typical responses to Malthus's population growth model are focused on technological innovations, potentially enabling mankind to overcome any economic growth constraints. For example, an appeal to the role of continuous technological progress is at the core of Marx's argument against Malthus. Marx recognizes the expansion of human needs throughout history but insists on the ability of the growing technology-based productive forces to satisfy them. In this spirit, Engels, as early as 1843, argued that even if the yield of agricultural productivity does not increase in proportion to the increase of the labor force (as a consequence of the population growth), "there still remains... science—whose progress is as unlimited and at least as rapid as that of population."[21] Overall, the unquestionable reliance on the unbounded expansion of science and technology has been characteristic of all technological optimists, independent of their political affiliation, from the times of Descartes.

In the twentieth century, Malthusian ideas took a more subtle form, supported by advanced statistical analysis and computerized modeling techniques. In the 1972 influential Neo-Malthusian bestseller *The Limits*

---

[20] Malthus, T. (2015)[1798]. *An essay on the Principle of Population.* Penguin.
[21] Engels, F. (2010). *Outlines of a Critique of Political Economy* (Vol. 3 p. 440). (In Marx and Engels. *Collected Works.* Lawrence & Wishard. Originally published in *German-French Annals (Deutsch-Französische Jahrbücher)* in 1844. Marx calls this work a "genius outline of the critique of economic categories."

to *Growth*, the authors use the "system dynamic theory and computer modeling to analyze the long-term causes and consequences of growth in the world's population and material economy."[22] The 1992 and 2004 updates of the original study address the criticism and further refine the model. Their quantitative analysis presents "possible patterns of world development ... from 1900 to 2100."[23] The central idea of the study is that the global economy has already entered into what they call the state of *overshoot*, defined as "the relation between humanity's demands on the planet and the globe's capacity to provide."[24] In this context, the authors refer to the research by Mathis Wakernagel and his colleagues, showing that as of 1999, the *ecological footprint of the world economy exceeded the earth's carrying capacity* by 20% of the available earth resources.[25] Though the overshoots are typical for any self-regulating system, depending on the extent of the overshoot, a system may either correct itself and return to the state of (dynamic) equilibrium or face a devastating collapse. All scenarios presented in *The Limits to Growth* indicate that to avoid global collapse, the current overshoot, caused by the sharp growth in global consumption, requires immediate corrective policy changes. The authors conclude that the objective of unbounded economic growth should be rejected and replaced by a *zero growth* model.

Summarizing critical responses, the authors point out that "the most common criticisms of the original... model were that it underestimated the power of technology" because *the value of natural resources is itself a function of the potentially unlimited technological growth.*[26] An appeal to science and technology, typical of technological optimists, can be traced at least to Fichte (see Chap. 2). Echoing Fichte, Julian Simon claims in *The State of Humanity* that due to unlimited human ingenuity, "the material conditions of life will continue to get better for most people, in most countries, most of the time, *indefinitely.*"[27] Likewise, Surrey and Broamley declare that "it would be unrealistic and probably dangerous to suppose

---

[22] Meadows, D., Randers, J. Meadows, D. (2004). *Limits to Growth: The 30-Year Update.* (p. IX). Chelsea Green Pub. For the application of systems dynamic approach, see Forrester, J. W. (1971). *World Dynamics.* Wright-Allen Press.

[23] Meadows et all, (2004), p. X.

[24] Meadows et all, (2004), p. 3.

[25] Meadows et all, (2004), p. XIV.

[26] Meadows et all, (2004), p. 204.

[27] Simon, J. L. (1996). *The State of* Humanity (p. 642). Wiley-Blackwell.

that the technological progress *could ever stop.*"²⁸ Such an attitude reflects the providential idea of progress deeply embedded in the European tradition. What all critics of the limited growth model share, independent of their political affiliation, is *blind faith* in the unlimited potential of science and technology.

The 2004 update of the *Limits to Growth* explicitly addresses the impact of technology on the 'doom and gloom' scenario. The authors explain that their model does not presume "technological progress at rates that would automatically solve all problems associated with exponential growth in the human ecological footprint" because technological innovations in a free market environment cannot be expected to provide *instantaneous and adequate* responses to ecological overshoots.²⁹ The core of this argument is of a socioeconomic nature and related to the interaction of technology with free markets: "Without signals from the market, the technology will not be forthcoming. Without technical ingenuity, the market's signals will produce no results."³⁰ In other words, in a free market economy already under the state of the 'overshoot,' the required policy responses would inevitably come with a significant delay, leaving the world economy in danger of (irreversible) collapse. The crux of the argument is that the modern markets are too complex to allow technology-based policy responses to take effect in a timely manner. Hence, the authors insist that a *zero growth* economic model is the only answer to the challenges of accelerating global economic expansion.

However, the *pragmatic* nature of the overshoot argument leaves a *theoretical* possibility that expanding human 'wants' are, in principle, capable of continuously triggering the revolutionary technological innovations that would allow—albeit with better planning—to reverse any adverse side effects of global growth, thus avoiding an ultimate collapse scenario. Then, what the Rome Club analysis may entail is the necessity to introduce or develop state-based social planning into the current market system, which is already in place, to a different degree, in all developed economies. Under such conditions, a *zero growth* policy could be used as a short- to medium-term measure, but the path of infinite growth would remain intact in the long run.

---

²⁸ Cole, H. S. (Ed.). (1973). *Models of Doom. A Critique of the Limits to Growth*. Universe Books.
²⁹ Meadows et all, (2004), pp. 203–204.
³⁰ Meadows et all, (2004), p. 208.

In an attempt to tackle the issue of technological growth from a strictly *theoretical* standpoint, Ridker, one of the *Limits* supporters, appeals to the familiar second law of thermodynamics. While Bogdanov uses the second law to prove *inexhaustible* human creativity, Ridker employs it as a theoretical justification of the *limited growth* model: "Technological breakthroughs may make it appear to be possible to continue growth forever. But this illusion arises from man's myopia... There is no such thing as a perpetual motion machine."[31] He seems to suggest that unbounded technological progress is equivalent to allowing the existence of the perpetual motion machine, something impossible due to the irreversible loss of useful energy and the accompanying increase of entropy. However, as was discussed above, the second law of thermodynamics applies only to a *closed system*, while in the *epistemologically open system*, the unlimited progress of theoretical and applied science (based on inexhaustible creativity) could, in principle, overcome any physical constraints.

In the last analysis, the arguments by the proponents of limited growth policies and their critics do not adequately address the principle of theoretically unlimited technological growth. The solution to the problem of *technological limits* depends on how the *limits to epistemology* are construed. In the framework of the Cartesian epistemological circle, this problem can be interpreted as a clash between the epistemic quest for infinite scientific pursuit (and its practical applications) and the ontological constraints of natural human existence. In the spirit of Kantian critique, we should examine whether or not the problem of infinite technological progress can be formulated in terms of the 'critique of scientific reason' and resolved in a purely *epistemic framework* independent of any naturalistic assumptions. In the hope of shedding light on this issue, in the next section, we explore the key arguments in the 'end of science' debate, focused on whether infinite progress of techno(science) is at least *theoretically* guaranteed by the open-ended nature of natural science or/and pure mathematics.

---

[31] Ridker, R. G. (1973, December). "To Grow or not to Grow" *Science*. Vol 182, 4119, pp. 1315–1318.

The End of Science and Invariants of Human Experience

The 'end of science' discussions have a long history, going back to at least the nineteenth century. Among the notable landmarks is an article by Emil Du Bois-Reymond, *The Limits of Our Knowledge of Nature* (1874).[32] The issue of scientific limits has repeatedly resurfaced over the last few decades, and a popular overview of the topic can be found in *The End of Science* (1996) by journalist John Horgan. The book is a collection of opinions by notable scientists on whether science, especially 'pure' science, could eventually encounter a 'saturation' stage, with any new theoretical advancements becoming, at best, incremental.[33]

Approaching the issue of scientific limits, we cannot overlook that the multifaceted nature of modern science is based on the historical "advance in instrumental design." From this perspective, the questions of limits to technology and science are intimately interconnected. As Don Ihde observes, the "use of technologies was already a part of the Renaissance lifeworld... Modern science is and has been essentially and historically technologically embodied." The technological embodiment of science has been accompanied by the increasing role of social institutions and the changing cultural milieu. In this way, 'pure' science took the form of what Bruno Latour dubbed 'technoscience,' developing in the instrumental, social, and cultural contexts.[34] In this frame of reference, the professed saturation of science is most often construed in the context of practical or *external* boundaries, such as social unrest, economic problems, or a shift in cultural attitudes. As summarized by Jurgen Mittelstrass, "the limits of science are either *error* limits or *economic* limits or *moral* limits," with the economic limits viewed as a chief, a quasi-theoretical decelerating factor making scientific advances increasingly difficult.[35]

---

[32] Du Bois-Reymond, E. (1874, May). "The Limits of Our Knowledge of Nature" In *Popular Science Monthly*, Vol. 5.

[33] Horgan, John. (1996). *The End of Science*. Broadway Books.

[34] Bruno Latour introduced the term of 'technoscience' In Latour, B. (1987). *Science in Action*. Harvard University Press. Don Ihde. (1991). *Instrumental Realism*. (p. 107). Indiana University Press. He cites the expression "advance in instrumental design" (p. 67) from Alfred North Whitehead (1963). *Science and the Modern World*. (p. 107). New American Library.

[35] Carrier, M., Massey, G. J. Ruetsche, I. (2000). *Science at Century's End* (p. 82). University of Pittsburgh Press.

Whether these factors may or may not lead to the 'end' of science, they do not imply any *intrinsic, epistemological* boundaries to scientific expansion. Addressing the issue of boundaries, Horgan points to an argument by a noted biologist, Gunther Stent, that "certain fields of science are limited by the boundedness of their subject matter."[36] Stent argues that biology is a bounded science, assuming (in the 1960th) that most of the major theoretical advances in biology are over, and "biologists had only three major questions to explore: how life began, how a single fertilized egg develops into a multicellular organism, and how the central nervous system processes information."[37] From this perspective, biology would be an example of an *ontologically* bounded science. Of course, one cannot overlook the paradigmatic nature of the 'only' questions left before the assumed 'end' of biology.

There is a subtle twist to Stent's concept of ontologically bounded discipline when he states that "bounded scientific discipline may well present a vast and practically inexhaustible number of events to study." At the same time, he argues that, though a *numerical* domain of "events to study" may be "practically" infinite, the truly *qualitative* growth would encounter a saturation stage of progressively diminishing returns, where "further efforts are of ever-decreasing significance." Thus, bounded disciplines are *qualitatively limited,* though the expected limits would be approached in gradual approximation rather than emerge as an abrupt barrier to any further advances.

While any *ontologically* bounded science is also *epistemologically* bounded, considering the limited domain of inquiry (as there will be nothing more to study), the situation with the unbounded sciences is more complex. Stent maintains that "there is at least one scientific discipline... which appears to be *open-ended*, namely physics, or the science of matter." Unlike biology, physics has no 'intrinsic limits' or clearly defined boundaries and can only be "expected to encounter limitations in practice"; the view that natural science may encounter practical limits but has no intrinsic boundaries has proponents among both philosophers and science practitioners.

Yet, some of the practical limitations to physics discussed by Stent seem borderline theoretical. In particular, he claims that "the very open-endedness of physics seems to be bringing it to heuristic limitation." He

---

[36] Stent, G. S. (1969). *The Coming of the Golden Age* (p. 111). Natural History Press.
[37] Stent, (1969), p. 10.

considers a situation where physics or some of its 'frontier disciplines' (such as cosmology and high-energy physics) are "moving rapidly toward a state in which it is becoming progressively less clear what it actually *is* that one is ultimately trying to find out." From this perspective, "the pursuit of open-ended science"—like the pursuit of a bounded science—"also seems to embody the point of diminishing intellectual return."[38] While in the case of 'bounded' biology, the diminishing return is an outcome of the diminishing subject matter (its structural complexity), for 'open-ended' physics, the alleged saturation is a matter of expected *erosion of goals*. The difference between the erosion of goals and the limitations of the subject matter is one between *epistemological* and *ontological* perspectives. Stent assumes that the complexity of the organic matter—the subject matter of biology—is ontologically 'bounded.' By contrast, the complexity of inorganic matter—the subject matter of physics—is unlimited; therefore, physics is an ontologically 'open-ended' but epistemologically bounded science due to the potential 'heuristic limitations.'

In fact, Stent's concerns about heuristic limitations raise a question of the ultimate *value* of our pursuit of truth, famously addressed by Nietzsche at the outset of *The Beyond Good and Evil:* "*What* really is this '*will to truth*' in us?" and what is "the *value* of this *will?*"[39] With a more pragmatic outlook on science, we could (along with Heidegger) consider science as applied technology instead of viewing technology as applied science.[40] From this perspective, the seemingly abstract constructs of physics—independent of their moral value and emotional appeal—have an *instrumental value* sufficient to motivate the continuing advancements of physics and contribute to open-ended technological progress. Then, the instrumental value of the practical (aka technological) applicability of science would be sufficient to secure its continuing progress. In such a value system, technological and scientific progress would also find support from a wide range of economic stimuli.

However, Stent addresses still another, strictly epistemological, constraint on physics as an open-ended science. Citing the late French physicist Pierre Auger, he stipulates "a possibility that there are mental limits to physics because of man's boundaries of intellect." In the spirit of

---

[38] Stent. (1969), pp. 112–113.

[39] Nietzsche, F. *Beyond Good and Evil* (Chap. I (1)).

[40] Heidegger, M. (1993). "The Question Concerning Technology" In Heidegger, M. Basic Writings. HarperCollins Pub.

epistemological naturalism, Auger and Stent consider such "mental" boundaries from a neurological perspective.

The neurological approach to the boundaries of intellect assumes that there is a correspondence of "a natural limit to the range of abstraction and complexity which can be covered by human thought" and the limitations arising from finite "number of nerve cells in the brain cells... and connections established between them." Stent sharpens Auger's argument, emphasizing that the limit in question "evolves from the circumstance that the fundamental... epistemological concepts, such as reality and causality, arise from" interaction between environment and "the genetically determined wiring diagram of our brain."[41] He reinforces this idea by attributing the origin of *a priori concepts* to the evolution of the human brain and traces this approach to Nietzsche and Konrad Lorentz, who "pointed [that] apparently miraculous concordance between intuition and the world can be easily explained by the theory of evolution."[42] This line of argument implies naturalistic (evolutionary) constraints on the ability of the human mind to proceed "from one essence to a deeper essence" (in Lenin's words) and resonates with Marx's ontology of the structural correspondence between human physical organization and nature (conceived as the human *extended body*). In the related line of thought, Marx underlines the relationship between (biology of) "the productive organs of man" and technology (as a "basis of all social organization") and raises a question of the evolutionary approach to technological progress. Under these conditions, the question of the infinite structural complexity of nature is epistemologically 'bracketed' and physics should be classified as *bounded* science.

Traditionally, mathematics plays a special role in philosophy and science. While Auger suggests that mathematical abstractions and concepts would fall under the neurological limitations in question, Stent claims that "mathematics belongs to a special category" and maintains that its open-ended nature is secured by Gödel's incompleteness theorem, with a reservation that mathematics also may "soon reach a point of diminishing returns."[43]

---

[41] Stent. (1969), p. 114.

[42] Stent, G. S. (1992). "Cognitive Limits and the End of Science." In *The End of science? Attack and Defense*. Nobel Conference XXV (R. Q. Elvee. Ed). University Press of America. The evolutionary interpretation of the a priori was already explicitly suggested by Herbert Spencer.

[43] Stent, (1969), pp. 113–114.

While Gödel's results were often interpreted in the sense that "mathematics was mysteriously open-ended," it is an open question whether such an interpretation is justifiable.[44] Gödel's results specifically address Gilbert's program to develop an axiomatic foundation for mathematics, with implications for the nature of mathematical *truth*. On this account, the limitations of incompleteness and undecidability he discovered are the limitations of a particular *interpretive framework* and do not imply unlimited and non-trivial 'progress' of mathematics as an 'unbounded' science.[45]

According to Gödel, any attempt at a *complete* axiomatic representation of the sufficiently rich subsystem of mathematics would be *inconsistent*, and any such *consistent* formal representation would be *incomplete*. Paraphrasing von Neumann's contention that the brain may be the only full description of itself, these results may indicate that mathematics is as well the only full description of itself. If we express this idea in Brower's intuitionistic framework, mathematics is *more doing than* being, and any attempt at its complete formal snapshot would be either inconsistent or incomplete. From this perspective, mathematics could be viewed as a form of *transcendental psychology* complementary to the naturalistic outlook of empirical science, and the limitations of the neurological 'diagram of the brain' would indeed match the 'a priori' limitations of the formalism of pure mathematics. Overall, whether framed (a la Kant) in the a priori, neurological or heuristic terms, constraints of the 'intellect' would represent theoretical limits to scientific progress.

Another attempt at a comprehensive analysis of scientific progress is presented by Rescher in the study, *The Limits of Science*, which received significant traction in the academic community.[46] His overall conclusion is that while there are no theoretical limits to scientific inquiry, science may have practical limitations that should not lead to either regret or

---

[44] Barrow, J. D. (1998). *Impossibility: The Limits of Science and the Science of Limits* (p. 210). Oxford University Press.

[45] Hilbert, in response to 'limits to science' arguments of Emil du Boise-Reymond, remarked: "In general, unsolvable problems don't exist... 'We must know. We will know'." Hilbert, D. (1935). *Gesammelte Abhandlungen* (3rd vol). Verlag Springer. By contrast, Hilbert's contemporary, Paul of du Boise-Reymond (brother of Emil) "showed himself a forceful critic of arithmetization and logicism and, in that respect... a direct ancestor of E. J. Brouwer." quoted from McCarty, D. C. (2013)."David Hilbert and Paul du Boise-Reymond: Limits and Ideals" (p. 6); oai: CiteSeerX.psu:10.1.1.366.4505 accessed at: https://citeseerx.ist.psu.edu/viewdoc/download?doi=10.1.1.366.4505&rep=rep 1&type=pdf

[46] Rescher, (1999). *The Limits of Science*. University of Pittsburgh Press.

complaint. Rescher assumes two types of such limitations: those based on technological shortcomings to meet the increasing scientific demands and those based on finite resources: "Given the constantly rising technological demands for continuing scientific progress, such advancement becomes increasingly more difficult and expensive in resource-cost terms. In the world of finite resources, this means that science must in the future progress ever more slowly—for strictly practical and ultimately economic reasons."[47] This position is not uncommon in the scientific community, particularly among physicists, who view the accumulation of practical constraints as a valid reason for the potential 'end of physics.'[48] For example, it is assumed that major advances in micro-physics will be hardly possible without the use of more and more powerful nuclear accelerators; accordingly, the inability to *infinitely* increase their power would impose practical constraints on nuclear physics.

As we argued above, the infinite qualitative progress of science and technology could, in principle, overcome *any* practical limitations, including resource limitations. On this account, practical constraints are only indicative of the *theoretically limited* potential of either cognitive (epistemic) depth or structural (ontic) complexity of nature. In this respect, Rescher keeps a gap between practical and epistemic constraints to scientific and technological progress in the same way that Bogdanov or the advocates of the 'limits to growth' argument do.

From a purely theoretical perspective, Rescher explores three lines of argument, addressing structural (ontic), operational (functional), and cognitive (epistemic) limits. Regarding the *ontological* line of argument, Rescher points out that in support of infinite scientific progress, "some theorists have felt compelled to stipulate an intrinsic infinitude in the structural makeup of the nature itself" and refers to the views of the noted physicist David Bohm.[49] It suffices to note here that the idea of "intrinsic infinitude... of the nature itself" represents a symbol of faith, returning us to the antiquated Leninist 'dialectic' of the "infinite process of deepening of human cognition of things, appearances, processes, etc., from appearance to essence and from less deep essence to deeper essence."[50]

---

[47] Rescher, (1999), p. 166.
[48] Feynman, R. (1994). *The Character of Physical Law*. Modern Library.
[49] Rescher, (1999), p. 78.
[50] Lenin, V. I. (1963). *Philosophical Notebooks* In Lenin, V. I. *Polnoye Sobranie Sochineniy* [Collected Works] (p. 203). Moskva: Gos. Izd. Polit. Lit.

At the same time, Rescher argues that the infinite scientific progress could be supported by "less radical appeal to... ever-deepening succession of operation principles or 'forces' within the physical makeup of nature."[51] While he refers, in this context, to the physicist Jean Paul Vigier, it is germane to note that Vigier assumes that "nature is infinitely complex."[52] Nevertheless, Rescher insists that "scientific progress actually" does not "require structural infinitude in the physical composition of nature," and what suffices is "functional" or "operational" complexity: "An unending depth in the *operational* or *functional complexity* of nature would be quite enough to underwrite the potential limitlessness of science." He explains that "even a mechanism of finitely ramified structure can have endlessly complex laws of operation" where "one can have rearrangements of items... ad infinitum with emergent lawful characteristic arising at every stage."[53]

Unfortunately (or fortunately), an issue of lawful characteristics of nature is not a matter of simple combinatorics because lawful characteristics do not arise from the arbitrary "rearrangement of items." Most of such 'rearrangements' would be useless or have no 'lawful characteristics' at all (especially "arising at every stage"). For instance, in reference to Rescher's examples of "letters... sentences... books," one could produce infinitely many semantically meaningless but syntactically well-formed statements (and books); even more, one may be able to produce an infinite number of meaningful expressions making *sense,* but only a finite number of such expressions would have a *reference* ('the king of France' has sense but no reference as France is not a monarchy). *If* the reference domain is infinite, *then* symbolic systems like mathematics or/and language may express this infinity, but formal systems do not generate an infinity of the related reference domains (in the extreme case, it could be an empty set).

Another version of the functional (operational) approach to scientific progress emerged, most recently, on the heels of the proliferation of Big Data datasets and advances in machine learning. In its radical form, this approach is presented by a popular thesis of the "end of theory," replacing the 'traditional' causal scientific explanations with Big Data ensembles and

---

[51] Rescher, (1999), p. 79.
[52] Rescher, (1999), p. 263 n. 29
[53] Rescher, (1999), pp. 80–81.

"correlations."[54] The end of theory contention represents a version of Rescher's operational complexity argument prone to similar criticism. In particular, many Big Data correlations would be as meaningless as Rescher's infinite "rearrangements of items." Calude and Longo emphasize this point, arguing that "the more data, the more arbitrary, meaningless and useless (for future action) correlations will be found in them. Thus, paradoxically, the more information we have, the more difficult is to extract meaning from it. Too much information tends to behave like very little information."[55]

In fact, the end of the theory argument returns us to a rather peculiar (and, in a sense, naïve) version of the Humian empiricism. Yet, even leaving alone the issue of causality, we can reiterate Kant's response to Hume: scientific knowledge is a product of *active* human interaction with the world (hence, a priori mathematical constructions constitute the only true science). In the case of data mining, the experts contribute to the emergence of Big Data ensembles and corresponding 'correlations' with irreplaceable human subject matter expertise, from problem definition to cleansing data, evaluating data noise and interpreting results. From this perspective, *the idea of the end of theory is as fictitious as that of the end of history.*

Following the long-standing tradition, Rescher views mathematics as a foundation of natural science. While he claims that the endlessness of natural science is *secured by the endlessness* of *pure mathematics*, the connection between the two is not as straightforward as he seems to assume. Moreover, the notion of *actual infinity* is controversial even within pure mathematics itself and is explicitly rejected by constructivist schools, including *intuitionism*. The applicability of actual infinity to nature (and natural science) is even more questionable, as mathematical models do not guarantee their empirical relevance.

In particular, the idea that potential infinity of mathematical becoming does not imply the actual infinity of nature goes back to the Aristotelian analysis of Zeno paradoxes. In the context of Hilbert's discussion of the first Zeno paradox, Kleene points to the "impossibility of

---

[54] Anderson, C. (2008, June 23). "The End of Theory: The Data Deluge Makes the Scientific Method Obsolete." *Wired.* https://www.wired.com/2008/06/pb-theory/

[55] Calude, C. S. & Longo, G. (2017). "The Deluge of Spurious Correlations in Big Data." *Foundations of Science*, 22(3), pp. 595–612. See also Kitchin, R. (2014, April). "Big Data, New Epistemologies and Paradigm Shift." *Big Data & Society*, 1(1), pp. 1–12. DOI:10.1177/2053951714528481.

drawing upon the perceptual or physical world" of abstract mathematical conceptualizations:

> We are by no means obliged to believe that the mathematical space-time representation of motion is physically significant for arbitrarily small space and time intervals; but rather have every basis to suppose that mathematical model extrapolates the facts of a certain realm of experience... similarly to the way the mechanics of continua completes an extrapolation in which a continuous filling of the space with matter is assumed... The situation is similar in all cases where one believes it possible to exhibit directly an [actual] infinity as given through experience or perception... Closer examination then shows that an infinity is actually not given to us at all, but is first interpolated or extrapolated through an intellectual process."[56]

The key point of this example is that the structure of the mathematical continuum cannot be directly applied to 'perceptual experience' and justify a claim of 'infinite structural complexity' of the material universe and, by extension, the unbounded nature of physics. In his 1925 talk, amidst the early twentieth-century scientific revolution, Hilbert points out that the infinity of mathematical continuum is not applicable in quantum physics and the infinity of Euclidian geometry—in Einstein's general relativity theory.

Hilbert underlines that even in pure mathematics, "in general must deductive methods based on the infinite be replaced by finite procedures which yield exactly the same results." This approach is at the foundation of his aspiration "to establish once and for all the certitude of mathematical methods." He emphatically states, "Our principal result is that the infinite is nowhere to be found in reality. It neither exists in nature nor provides a legitimate basis for rational thought—a remarkable harmony between being and thought."[57] In this context, neither the universality of mathematics nor natural science requires the substantive concept of the infinite. Infinity is rather a regulative idea of (mathematical) reason to be utilized within a well-founded axiomatic system.

---

[56] According to Zeno's first paradox, a runner cannot cover a distance in a finite time because the course is infinitely divisible continuum. Hilbert, D. and Bernays, P. (1934). *Grundlagen der Mathematik*. Springer. Quoted from Kleene, S. C. (1971). *Introduction to Metamathematics* (p. 55). Wolters-Noordhoff Pub.

[57] Hilbert, D. (1926). "On the Infinite," *Mathematischen Annalen*, vol. 95, pp. 161–190. https://doi.org/10.1017/CBO9781139171519.010 (pp. 184, 201).

The decisive, for Rescher, *epistemological* line of reasoning is supposed to supersede those already discussed. It is based on the claim that the intrinsic infinity of science is guaranteed by the infinite "conceptual depth" ('inexhaustible creativity' in Bogdanov's lingo). Says Rescher: "In fact, however, neither unending structural nor operational complexity is required to provide for cognitive inexhaustibility" and, consequently, the "potential endlessness of scientific progress."[58] Elaborating on this argument, Rescher correctly states that "scientific progress hinges not just on the structure of nature itself but also on the structure of the information-acquiring process by which we investigate it." He continues, "Such an approach to the infinity of nature construes this infinitude not so much in terms of spatial extensiveness or physical structure as of *conceptual depth*" and summarizes this position by stating that "the crux is not nature itself, but how we explorers manage to deal with it. Responsibility for the open-endedness of science need not lie on the side of nature at all but can rest *one-sidedly* with us, its explorers."[59]

In the footnote, Rescher observes that "nature-as-it-is cannot... be simpler than nature-as-it-is-thought-to-be, because the latter is... a part of the former."[60] In other words, if man is a part of nature, the structural complexity of nature cannot be "simpler" than (belonging to it) cognitive apparatus. While this is a perfectly valid statement, it does not entail that either human cognitive apparatus or nature is infinite in any conceivable sense. If man is a "part" of nature, as Rescher recognizes, his cognitive apparatus is finite and obviously simpler than "nature-as-it-is" unless we subscribe to the Cartesian dualism of two substances. As Stent, Auger, and Lorentz suggest, human neurological makeup imposes limits on cognitive exploration of nature exactly because the human brain is only a part of nature. Then, the abilities of "us, its explorers" cannot *one-sidedly* guarantee the infinite 'open-endedness' of science. In fact, Rescher's analysis is a step back even from the balanced anthropocentric ontology of Marx's 1844 *Paris Manuscripts*. While Rescher's *one-sided* and highly speculative claim that "Nature herself has no 'depth'" seemingly echoes Marx's view that "nature too, taken abstractly, for itself, and rigidly separated from

---

[58] Rescher, (1999), p. 81.
[59] Rescher, (1999), p. 66.
[60] Rescher, (1999), p. 263 n. 32

man, is nothing, for man," for Marx, scientific and technological advancements appear as a product of the complex interplay between the human physical organization and nature.[61] In this context, the isomorphism between subjective and objective nature does not involve the (actual) infinity of either of them.

From this perspective, the intrinsic correspondence between the human perceptual and cognitive apparatus and nature renders the assumption of 'inexhaustible' creativity or structural complexity of nature 'in itself,' at best, *redundant* for the universality of man and his scientific pursuit. By the same token, redundant is the thesis of 'absolute' rationality applicable to 'any rational being.' A system of knowledge could be defined only relative to a particular mode of instrumental/cognitive activity. There is no *ontologically fixed* point from which the scientific and instrumental appropriation of the world could proceed. As Bachelard emphasizes, the absolute reason does not exist and "'rationalism' need not be a closed system."[62] The apparently immutable logos of the universe refers to stable, *invariant* relations emerging in historically unfolding human instrumental activity. The *invariants of experience* characterize human experience rather than an abstract 'reality in itself' or Platonic essences waiting to be grasped by any rational being.[63] In this framework, while the singularities of mundane experience represent the 'thingness' or 'invariants of perception,' the scientific 'objects' represent the *invariants of instrumental action*.[64] The 'infinity in itself,' either in the form of structural complexity of nature or inexhaustible creativity, has no place in this picture of the world.

---

[61] Rescher, (1999), p. 82. Marx, K. (1971). *Economic and Philosophical Manuscripts*. In: Fromm, E. *Marx's Concept of Man with a Translation from Marx's Economic and Philosophical Manuscripts* (p. 183) (T. B. Bottomore. Trans.). Frederick Ungar Pub.

[62] Bachelard, G. (1984). The *New Scientific Spirit (p. 3) (A. Goldhammer. Trans)*. Beacon Press.

[63] Consider a related Cassirer's position in Cassirer, E. (1923). *Substance and Function and Einstein's Theory of Relativity*, pp. 265–270. Though the epistemological point of reference for Cassirer is Kant's transcendentalism, his ideas are also marked by Mach's influence.

[64] Max Born articulates this position in "Memories and proceedings of the Manchester Literary and Philosophical Society" (1949–1950). Vol. 91. It is closely related to the views of Copenhagen school, including Niels Bohr and Werner Heisenberg.

Coda: Knowledge, Freedom, and the Riddle of History

The arguments of either the infinite structural complexity of nature or the inexhaustible human creativity represent two poles of the historically recurrent Cartesian circle that, in Koyre's words, "characterizes... any epistemology, because any epistemology is intended to discover both the objective sense and the [subjective] means of knowledge. The circle is contained in any act in which knowledge has itself as an object. In this self-reference, knowledge touches the Absolute."[65] Descartes attempts to fill the gap between the means and object of knowledge by viewing mathematics as a model of rational thinking and nature as a mathematical manifold to be theoretically appropriated by science and practically—by technology. In the Cartesian scientific paradigm, mathematics provides a universal framework, allegedly allowing man to overcome the dualism of spirit and matter and 'touch the Absolute' in the strife for the ever-growing mastery over nature.

At the dawn of the twentieth-century scientific revolution, the Cartesian ideal of a rationally organized *objective reality* revealed, layer after layer, by natural science, was explicitly articulated by Max Planck, who conceptualized scientific progress as a gradual approximation towards the picture of the measurable universe, accessible to any thinking being and purified from 'anthropomorphic' elements of mundane experience.[66] Generalizing the spirit of modernity, Husserl says, "In the bold... elevation of the meaning of universality, begun by Descartes, this new philosophy seeks nothing less than encompass, in the unity of a theoretical system, all meaningful questions in a rigorous scientific manner... in an unending but rationally ordered progress of inquiry," becoming "one edifice of definitive, theoretically interrelated truths," intended "to solve all conceivable problems—problems of facts and reason, problems of temporality and eternity."[67]

The paradigm of rigid Cartesian rationalism was decidedly undermined by Kant's epistemological critique, which construes objective reality as a product of the complex synthesis of external stimuli and constructive

---

[65] Koyrè, A. (1923). *Descartes und die Scholastic* (s. 97). Bonn
[66] Plank, M. *Die Einheit des Physikalischen Weltbild.* This lecture delivered in 1909 and reprinted in Plank, M. (1958). *Physikalische Abhandlungen und Vortage*, bd. 3, s. 6–29.
[67] Husserl, E. (1970). *The Crisis of European sciences and Transcendental Phenomenology* (pp. 8–9). Northwestern University Press.

activity of theoretical reason. The twentieth-century scientific revolution only deepened the crisis in the long-standing scientific practices. With the revolutionary advancements of twentieth-century science, Kantian apriorism takes a more sophisticated form in both pure mathematics and mathematical physics. Now, mathematics, as an instrument of rational investigation, resorts to the science of structures (such as groups, rings, etc.), while "every occupation with structures... presupposes an intuitive operation with natural numbers or equivalent operation with symbols."[68] Elaborating on these developments in the spirit of Kant, Poincarè says that the "general concept of group preexists in our mind, at least potentially. It is imposed on us... as a form of our understanding; only, from among all possible groups, we must choose one... to which we shall refer natural phenomena." He underlines that such choice is a matter of *convenience* (hence, his 'conventionalism').[69]

With the foundational principles of mathematics increasingly penetrating the conceptual underbelly of other fields, the invariant (group) relations are discovered in mundane sensory experience and its extension perpetuated by technoscience. In this context, Cassirer refers to Helmholz's early attempt to apply the notion of a group to psychology and suggests that "the application of concepts of this type extend both farther and deeper... to the very roots of perception itself."[70] At the core of these developments is our ability to express the perceptual experience in algebraic terms and relate it to the constructive activity of reason by choosing *invariants* "to which we shall refer natural phenomena." In the last analysis, at stake are the foundations of the unity between mind and matter pertaining to the very core of human universality.

Bachelard addresses the radical shift in the epistemological paradigm in the context of Heisenberg's uncertainty principle: "Early in the development of modern science, Descartes argued that it was important to explain natural phenomena in terms of figures and motions. The uncertainty relations express the fact that a rigorous description of this kind is impossible

---

[68] Ackermann, W. (1957, December). "Philosophical Observations on Mathematical Logic and on Investigation into the Foundations of Mathematics." *Ratio*, I, 1, p. 12.

[69] Poincare, H. (1905). *Science and Hypothesis* (p. 82). The Walter Scott Pub.

[70] See Cassirer, E. (1944) "The Concept of Group in the Theory of Perception." *Philosophy and Phenomenological Research*, 5, 1. pp. 1–35. Cassirer tries to reconcile the elements of Kantian apriorism with Planck's program of eliminating the anthropomorphic elements from science.

since we can never know the figure and its motion *at the same time*."[71] Bachelard concludes that "the uncertainty relations should be interpreted as impediments for absolute analysis. In other words, we must define the basic notions of physics in terms of relations, just as we define mathematical objects by stating the axioms that determine how they relate to one another. Parallel lines do not exist *prior* to Euclidean postulate; they are *ulterior*... Definitions depend on methodology: Tell me how to find you and I will tell you what you are."[72] Then, each mode of scientific exploration constitutes its own domain of being, and the growing tree of 'techno-science' is defined through (institutionalized) paradigms evolving in human practice.

Emphasizing the pragmatic nature of science, Herman Weyl, Einstein's Princeton colleague and a contributor to the twentieth-century scientific revolution, underlines that science "simply endeavors to prolong a certain important line already laid out in the structure of our practical world"—a genuine medium of scientific practice.[73] From this perspective, a *mode of being* and the rational *means of knowledge* are the two inseparable moments of scientific pursuit. The act of knowledge could be compared to tossing a stone into the body of still water: the objectivity of ripples would be inseparable from the activity of the knower; asking what the ripples are 'in themselves' is akin to a famous Zen *koan*: What is the sound of one hand's clap? This approach implies that the knower is an intrinsic part of reality; therefore, cognition represents a complex relationship between subjective information processing, sensory apparatus, and (using Marx's expression) the extended body of nature. On this account, the infinite (in any nontrivial sense) penetration into the 'deeper and deeper essences' could be justified only within the antiquated Cartesian dualism or Fichtean radical subjectivism, with man 'one-sidedly' defining infinite scientific progress.

Considering the long-standing role of mathematics as a "bond between man and the universe" and a "key to future understanding of the cosmic and moral order," it is not surprising that the twentieth-century revolutionary changes in natural science and philosophy found an expression in the paradigmatic shift in the foundations of mathematics.[74] Instead of

---

[71] De Broglie is cited by Bachelard from de Broglie, L. (1932). *Theorie de la Quantification dans la Nouvelle Mecanique* (p. 31). Paris.

[72] Bachelard, G. (1984). *The New Scientific Spirit*. Beacon Press, p. 138–139.

[73] Weyl, H. (ca 1949). "Man and the Foundations of Science" In Weyl, H. (2009). *Mind and Nature*, p. 188.

[74] Cassirer, E. (1953). *An Essay on Man* (pp. 32, 34). Doubleday and Co.

being 'grasped' in their ideal purity of the Platonic universe, mathematical entities are conceived as dynamic structures, constituted through the medium of 'pure' or symbolic construction. In Brouwer's radical constructivism, they are a product of the spontaneous synthesis of volition and reason (referred to by Kant as productive imagination). Brouwer construes *time* as an innermost medium of constructive activity and a fundamental constitutive element of human experience: "Consciousness in its deepest home seems to oscillate slowly, will-lessly, and reversibly between stillness and sensation. And it seems that only the status of sensation allows the initial phenomenon of the said transition. By a **move of time** a present sensation gives way to another present sensation in such a way that consciousness retains the former one as a past sensation, and moreover, through this distinction between present and past, recedes from both and from stillness, and becomes **mind**."[75] Experience, in all its modes, is *invariant* to the "move of time."

The initial act of freedom, expressed in the transcendental 'oscillation' of time, represents a sort of intuitive Big Bang, a foundational mode of the *a priori synthesis*, establishing the scale of phenomenal time series and, thus, unfolding subjective experience into the manifold of 'objective world.' This ontology of freedom demonstrates the complementarity of acting and being, merging into an inseparable unity within the immanency of human subjectivity. The time-based constructive activity and its static, retrospective snapshot of *being* are the two complementary aspects of a unitary act of free self-positing of the Ego. Weyl pointedly remarks that "with [constructivist] mathematics we stand precisely at that intersection of bondage and freedom that is the essence of the human itself."

In fact, Brouwer's transcendentalism goes well beyond mathematics itself (alienating some of his followers) and not only 'stands at' but decidedly crosses "the intersection of bondage and freedom," penetrating the existential core of human subjectivity.[76] The problematic of existential

---

[75] Brouwer (2011)[1981]. "Philosophy, Consciousness and Mathematics" In van Dallen, D. (Ed.). *Brouewr's Cambridge Lectures on Intuitionism*. Cambridge University Press. For the account of key Brouwer's concepts of free choice sequence, spread etc. see Heyting, A. (1956). *Intuitionism. An Introduction*. North-Holland Pub. See also the section on Fichte in Chap. 2.

[76] Weyl, H. (2012). "Levels of Infinity" In Weyl, H. *Levels of Infinity*. Dover Pub. Weyl tries to combine intuitionism with the formalism of Hilbert. The more relaxed versions of constructivism utilize constructive (algorithmic) procedures without Brouwer's intuitive construction and the corresponding ontological commitments.

constructivism goes back to the crisis of Cartesian rationalism arising at the height of German transcendental tradition and resonates with the fundamental intuition of Schelling: "Man's actions are in principle predictable and that they are free. They are necessary; <u>but this necessity is an inner necessity, imposed by the ego's original choice</u>, not a necessity externally imposed by God.' This inner necessity is itself freedom, the essence of man is essentially **his own act**; necessity and freedom are mutually immanent...'"[77]

These existentialist reflections find a new voice in the twentieth-century metaphysics of freedom and necessity, time and being. Heidegger tries to link the initial act of self-positing to the ontology of Being: "Understanding of Being is itself a definite characteristic of Dasein's Being."[78] Characterizing the ontological openness of Being, Heidegger echoes Hegel, referring to human being-in-the world (*Dasein*) as "being held [in its relation] into the nothingness." The existential chasm of human being-in-the-world (*Dasein*) is open to the depth of Being through the "dealings in the world and with entities within-the-world," "lightening" the emerging circle of Being *for man* and giving it the meaning and unity of what is *real*. In this distinctly human relatedness to Being, modern man is getting engaged in the technological attitude of *enframing* (*Ge-stell*) his lifeworld, where, in Hegel's words, technology is getting increasingly 'higher' and 'more honorable' than the products of labor "it procures and which are the purposes."

Essentially, this lifeworld of enframing has been redefining "what it means to be a thing" and what it means to be human.[79] In this alienated mode of existence, technological practice is "driving on to the maximum yield at the minimum expense," creating a culture focused on reproducing technological artifacts and know-how. Built into the fabric of human life, these practices are overpowering both producers and consumers, transforming man into his externalized alter ego, entrapped in the infinite cycle of production and consumption. In such a technology-driven economic environment, man plays the role of on-demand "standing reserve," along with technology and nature as it emerges for man in technological practice.[80]

---

[77] Copleston, S. J. F. (1963–1965). *A History of Philosophy (Vol. 7, 2, p. 165)*. Image Books
[78] Heidegger, M. (1962). *Being and Time*. Harper and Row.
[79] Hegel, G. W. F. (2010). *The Science of Logic* (pp. 657–653). Cambridge University Press
[80] Heidegger, M. (1993). "The Question Concerning Technology" In Heidegger, M. *Basic Writings*. (pp. 311–341). Harper Collins. In his example, "The hydro-electric plant is not built into the Rhine River... Rather, river is dammed up into the power plant." The river, nature and man himself are always standing-ready for technological use.

The attitude of standing reserve increasingly reshapes popular culture through the design and use of disposable technological artifacts. Illustrating the contrast between the culture of standing reserve and the holistic lifestyle of traditional cultures, Dreyfus points to the difference between the disposable Styrofoam cup and the porcelain cup used in the Japanese tea ceremony. The latter symbolizes a tradition passed from generation to generation and is valued and preserved "for its beauty and its social meaning." In this context, he emphasizes the intrinsic connection between the self-awareness of being human and the understanding of "what it means to be a thing."[81]

The Styrofoam cup represents a design trend typical for the environment of standing reserve, with the designer, in a sense, 'alienated' from the design choices and turned into a vehicle of global market forces instead of exercising his 'essential powers' (in Marx's terms). The advances in the IT industry provide even more poignant examples of the growing 'disposable resource' culture. For example, the usage of hard-wired batteries and microchips (such as processors) in common consumer gadgets forces consumers to discard and replace them in a way similar to the used Storyfoam cups or paper bags. In a related development, the increasing memory and processing power are coupled with the 'liberal' software design, enhancing the trend of continuous rotation of consumer technology, with more and more products turning into disposable resources.

Moreover, the proliferation of global communication tools, Big Data mining and AI only amplifies the dangers of the persistent culture of technological determinism, providing new forms of technical control, making corporate giants 'too big to fail' and too dangerous to be allowed to run amok. While free from some of the ills of industrialization, recent advances in AI introduce a new set of problems, thus providing additional support to limited growth policies, focused (in Kant's words) on the culture of discipline rather than the utilitarian culture of skill. From this perspective, Feenberg justifiably points to the connection between the 'substantivist' understanding of technology and its role in "capitalism… freed to extend

---

[81] Dreyfus, H. L. (1995). "Heidegger on Gaining a Free Relation to Technology" In A. Feenberg and A. Hannay (Eds.). (pp. 97–108). *Technology and the Politics of Knowledge*. Indiana University Press. Consider also the phenomenon of 'hostile architecture' in places like NYC, intentionally turning public areas, such as parks and subway stations, into usable resources rather than living spaces. One may say, we do not 'live' in these places, but we do! We live in the city as we live in our homes and country. The point is that architectural design should enhance living rather than turn a city into a standing reserve.

technical control to the labor force [and] the organization of work," with technology becoming an increasingly autonomous force of social alienation.[82]

The instructive illustration of an alternative, socially driven technical design practice is free, open-source software development, including the election of project design leaders, community-based testing and intensive communication between designers and users; in this distinctly social "form of objectivity in which technology reveals itself," the choice of technology involves a choice of social communication model behind its design, testing, and distribution. In this context, public support for open-source, free software design and use becomes increasingly important. Unlike commercial software, free software is not only democratic in its origin but more affordable and minimalistic in terms of hardware requirements.

Socioeconomic democratization involving technology exemplifies a path of individual and communal emancipation with far-reaching consequences, and it can find strong support in global multiculturalism. By contrast with the radical European rationalism, the holistic view of man and minimalistic approach to life have remained the vital elements of Eastern tradition. At its core is a conviction that the path to human universality, either individual or historical, lies through the humanistic synthesis of science and social practice, humanity and its natural environment.[83] Through the union of human activity and the creative power of nature, the latter emerges as, in Marx's words, an extended human body rather than a hostile material to be 'appropriated' by man as its 'master and possessor.' In this worldview, the genuinely humane historical telos is not to *dominate* nature through the unbounded expansion of technological control but to *adapt and grow into* nature, adjusting human existence to the natural Eco-system to "the unlimited extent." At the same time, unity

---

[82] Feenberg, A. "From Essentialism to Constructivism: Philosophy of Technology at the Crossroads." https://web.archive.org/web/20061205020238/http://www-rohan.sdsu.edu/faculty/feenberg/talk4.html. Criticizing Heidegger and Habermas (among others), Feenberg attempts to develop a 'broader,' historically contextualized view of technology. He claims that substantivism (essentialism) neglects "the social nexus" of "technical rationality." He elaborates on these issues in Feenberg, A. (2002). *Transcending Technology* (2nd ed). Oxford University Press.

[83] The essential element of major world religions and mythologies is an intimate relatedness between the sacred and profane. For example, in the Vedas, the creation of the world is a cosmic sacrifice, and an individual is to emulate it on the personal level; in the Gita, Krishna directs Arjuna to imitate his *karma yoga* path.

with nature would allow people to discover, develop, and exercise their uniquely human 'essential powers,' in the *modus operandi*, and reveal themselves in the universality of their species-life.[84]

This cultural attitude is well captured in E. F. Schumacher's remark that "the Buddhist sees the essence of civilization not in multiplication of wants but in purification of human character."[85] Characterizing what could be called a practice of *social karma yoga*, he reportedly remarks that "economics without... human and ecological values is like sex without love" and sees a viable alternative to uncontrolled Western consumerism in 'Buddhist economics,' where cultural values retain their predominant role in socioeconomic planning. The 'enlightenment' (pun intended), instead of being mediated by the alleged omnipower of science and technology, is to be achieved in a 'disciplined' way of 'talking' to nature and satisfying the wants sufficient to live a genuinely human, spiritual as well as natural, life. While a chief socioeconomic objective for an 'affluent society' is to obtain the maximum production with the minimum of work, the objective for the advancement of a humane 'culture of discipline' is "to obtain the maximum of well-being with the minimum of consumption." A proper balance implies that "it is not wealth that stands in the way of individual self-fulfillment but the attachment to wealth; not the enjoyment of pleasurable things but craving for them."[86] At the core of such outlook are the three pillars of Oriental philosophy: the principle of non-attachment to the fruits of one's actions, the conception of non-violence (*ahimsa*), and the focus on *unity* with all sentient beings and nature.

While such philosophy of life, at the personal and social levels, apparently runs in the face of the ideal of man as a 'master and possessor of nature,' it is by no means foreign to the long-standing Western tradition, going back at least to Spinoza's aspiration to direct sciences to moral perfection and Kant's appeal to uphold the humane culture of discipline over the instrumental culture of skill. In a similar vein, G. E. Lessing preaches that a genuine historical *telos* is a practice of 'education of the human race'

---

[84] By contrast, an instrumental attitude to nature is a way to treat it as "accidentally present and arbitrarily usable conglomeration of lifeless matter," a passive object of human activity; see Rapp, F. (1981). *Analytical Philosophy of Technology*. (p. 102). Dordrecht: D. Reidel Pub. A related discussion can be found in Eliade, M. (1957). *Das Heilige und das Profane*. Hamburg: Rowohlts Taschenbuchverlag.

[85] Schumacher, E. F. (1999). *Small is Beautiful: Economics as if People Mattered* (pp. 38–39). Harley & Marks.

[86] Schumacher (1999), pp. 40–41.

rather than the indulgence of senses in the ever-growing net of material wants. The same spirit is conspicuous in Marx's early humanism of social yoga, which is focused on emancipating humanity from the *sense of having* and establishing a direct relationship with nature through the development of all human senses; towards the end of his career, this attitude was further reinforced by his attention to the environmental impact of capitalism.

On this background, the crisis of Western historical universalism should not be viewed as a 'clash' of competing civilizations but as an opportunity, the opening towards the synthesis of Western cultural experience with that of other great civilizations; if the objective is a truly universal 'world history' rather than political, economic, or military supremacy, then globalization no longer should be viewed as identical to Westernization. To paraphrase a popular motto, not all Western values may be important, and not all the important values may be Western. In the context of Bergson's distinction between *closed* and *open* society, we should be careful to define what constitutes a society open to cultural and socioeconomic diversity and distinguish between historical universalism based on global supremacy vs. that based on deepening global multiculturalism.[87] In this context, the global expansion of technical rationality should not be mistaken for the development of a multicultural community of nations.

In the spirit of the philosophy of a 'practical idealist' Mahatma Gandhi, as three hundred million Indians stood up to the power of a hundred thousand Englishmen, the hundreds of millions of 'consumers' worldwide can stand up to the power of corporate power elites; and if seventy percent of the world's largest economy comprises consumer spending, then a personal choice *to be* or *to have* becomes a decisive factor of radical changes in the socioeconomic infrastructure and the whole fabric of human life. The change begins with one's commitment to non-violence and the discipline of minimum consumption sufficient for revealing one's potentialities. These objectives are inconsistent with the fetish of the unceasing technological self-externalization and the ever-growing new Babylonian Tower of human 'wants,' which is only substituting human essential nature with its technological alter ego. It is important to remember that the separation between cultural values and technology-driven socioeconomic environment is as questionable as a 'liberal' separation between human inner and outer life.

---

[87] Bergson, H. (1935). *The Two Sources of Morality and Religion*. MacMillan.

It is germane to note that the holistic practices of traditional cultures are closely related to artistic experience. While technological artifacts are defined by their serviceability in the predefined world of standing reserve, in artistic forms of objectification the work of art symbolizes a (contextual) disclosure ('clearing') of Being. By contrast with the potentially poignant nature of technological self-objectification, artistic objectification is rooted in the productive imagination of an artist, triggering the reciprocal catharsis of a spectator.[88] In this context, art should not be viewed as a commodity but as a way of 'species life,' essential not only for personal self-fulfillment but also for our ability to maintain and enhance a sustainable socioeconomic and political environment, where the 'outer' and 'inner' life find genuine balance.

Of course, the technological form of being has its own history, and man cannot 'exit' it at will. However, the critical reflection on the productivist socioeconomic model and the communal educational as well as political involvement have the power of *clearing* the horizon of human lifeworld, giving people an impetus to overcome the utilitarian approach to life of the on-demand, disposable resource. For an individual, such *clearing* provides a new direction in one's life, for humanity—in its historical destiny. The true infinity of human experience is rooted in the open abyss of Being and could be achieved only through free self-positing at the personal and community levels: if I am not for myself, then who is for me? And if I am only for myself, who am I? And if not now, then when?[89] It is through free personal choices that the phenomenon of (historical) time is being born, thus creating its own ontological foundation, the scope, and extent of possible experience, whether personal, communal, or global.

---

[88] Cassirer discusses this theme in the "Form and Technology." In *Ernst Cassirer on Form and Technology* (2012). (A. S. Hoel and I. Folkvord. Eds.). Palgrave Macmillan. Heidegger explicitly addresses the concept of *clearing* in the *Origins of Art*. In his diaries, Kafka refers to the writer's ability to silence one's soul and listen to the world opening for him.

[89] Rabbi Hillel. *Mishna: Pirkei Avot*, 13.

Bibliography

Adams, H. (1965). *Karl Marx in his Earlier Writings*. Russel & Russel.
Ackermann, W. (1957, December). Philosophical Observations on Mathematical Logic and on Investigation into the Foundations of Mathematics. *Ratio, I*(1), 12.
Adler, M. (1978). A Critique of Othmar Spann's Sociology (Zur Kritik der Sociologie Othmar Spann) (1927). In *Austro-Marxism*. (T. Bottomore & R. Goode, Trans.). Clarendon Press.
Althusser, L. (1970). *Lenin and Philosophy* (B. Brewster, Trans.). Vintage Books.
Anderson, C. (2008, June 23). The End of Theory: The Data Deluge Makes the Scientific Method Obsolete. *Wired*. https://www.wired.com/2008/06/pb-theory/
Anderson, K. (1995). *Lenin, Hegel, and Western Marxism: A Critical Study*. University of Illinois Press.
Aristotle. (1921). Politics. In *The Works of Aristotle* (Vol. 10, W. D. Ross, Ed., Jowett, Trans.). Clarendon Press.
Arthur, C. J. (1986). *Dialectics of Labor*. Basil Blackwell.
Augustine. (2008). *Confessions* (H. Chadwick, Trans.). Oxford University Press.
Avineri, S. (1978). *The Social and Political Thought of Karl Marx*. Cambridge University Press.
Бажанов, Б. Г. (1992)[1930]. *Воспоминания Бывшего Секретаря Сталина* [Bazhanov, B. G. The Memoirs of Former Stalin Secretary]. Ст. Петербург: Всемирное Слово.
Bachelard, G. (1984)[1934]. *The New Scientific Spirit* (p. 3, A. Goldhammer, Trans). Beacon.

Ballestrem, K. (1969, December). Lenin and Bogdanov. *Studies in Soviet Thought*, *9*, 283–310.
Barth, H. (1963). *Wahreit und Ideologie*. Manesse.
Baron, S. H. (1963). *Plekhanov. The Father of Russian Marxism*. Stanford University Press.
Barrow, J. D. (1998). *Impossibility: The Limits of Science and the Science of Limits*. Oxford University Press.
Bauer, B. (1841-1842). *Kritik der evangelischen Geschichte der Synoptiker* (3 Vols.). Leipzig.
Bauer, B. (1841). The Trumpet of the Last Judgment over Hegel the Atheist and Antichrist: An Ultimatum (Die Posaune des jüngsten Gerichts über Hegel den Atheisten und Antichristen: ein Ultimatum); The Genus and the Crowd. In L. S. Stepelevich (Ed.), (1983a). *The Young Hegelians: An Anthology*. Cambridge University Press.
Bauer, B. (1983). The Genus and the Crowd (1844). In L. S. Stepelevich (Ed.), (1983). *The Young Hegelians: An Anthology*. Cambridge University Press.
Bello, R. E. (1985, August). The Systems Approach. *Studies in Soviet Thought*, *30*, 131–148.
Bergson, H. (1935). *The Two Sources of Morality and Religion*. Macmillan.
Berlin, I. (1959). *The Life and Opinions of Moses Hess*. Cambridge.
Berlin, I. (1963). *Karl Marx* (3rd ed.). Oxford University Press.
Bernstein, E. (1899). *Voraussetzungen des Sozialismus and die Aufgaben der Sozialdemokratie* (Translated as Bernstein, E. (1993). *The Preconditions for Socialism*). Cambridge University Press.
Bernstein, E. (2015). *Evolutionary Socialism*. Forgotten Books.
Bernstein, R. J. (1971). *Praxis and Action*. University of Pennsylvania Press.
Bertalanffy, L. (1969). *General System Theory*. George Braziller.
Bogdanov, A. (1904-1906). *Empiriomonism*. Dorvatovskiy i Charushnikov.
Bogdanov, A. (1908). Ernst Mach und die Revolution. *Die Neue Zeit*, *19*.
Bogdanov, A. (1910). *Padeniye Velikogo Fetishizma. Vera i Nauka* [The Fall of Great Fetishism. Belief and Science]. Dorvatofskii i Charushnikov.
Bogdanov, A. (1913a). *Filosofia z'ivogo opyta* [Philosophy of Vital Experience] (2nd ed.: 1921, 3rd ed.: 1923].
Bogdanov, A. (1913b). *Tektologiya: Vseobschaya Organizatzionnaya Nauka* (Vol. 1). Semenov.
Bogdanov, A. (1917). *Tektologiya* (Vol. 2). Kniznoye Izdatel'stvo Pisateley v Moskve.
Bogdanov, A. (1922). *Tektologiya* (Vol. 3). Grez'bin.
Bogdanov, A. (1925-1929). *Tektologiya* (3rd ed.). Kniga.
Bogdanov, A. (1989). *Tektologiya* (Vol. 2). Economica.
Bogdanov, A. (1923, April). Printzip Otnositel'nosti i ego Filosofskoye Istolkovaniye [The Principle of Relativity and its Philosophical Interpretation]. *Mir*, 4.

Bochenski, J. M. (1963). *Soviet Russian Dialectical Materialism* (N. Sollohub, Trans.). D. Reidel.
Bohr, N. (1934). *Atomic Theory and the Description of Nature.* Cambridge University Press.
Bologh, R. W. (1979). *Dialectical Phenomenology—Marx's Method.* Routledge and Kegan Paul.
Broglie, L. de. (1932). *Theorie de la Quantification dans la Nouvelle Mecanique.* Paris.
Brouwer, L. E. J. (2011)[1981]. Consciousness, Philosophy and Mathematics In D. van Dallen (Ed.), *Brouewr's Cambridge Lectures on Intuitionism.* Cambridge University Press.
Brazill, W. J. (1970). *The Young Hegelians.* Yale University Press.
Bury, J. B. (1920). *The Idea of Progress: An Inquiry into its origin and Growth.* Macmillan and.
Cahoone, L. E. (1995, July). The Plurality of Philosophical Ends: 'Episteme, Praxis, Poiesis'. *Metaphilosophy, 26*(3), 220–229.
Callinicos, A. (1996, Jan-Feb). Messianic Ruminations: Derrida, Stirner and Marx. *Radical Philosophy, 75*, 37–41.
Calude, C. S., & Longo, G. (2017). The Deluge of Spurious Correlations in Big Data. *Foundations of Science, 22*(3), 595–612.
Cassirer, E. (1902). Leibnitz's System in Seinen Wissenschaftlichen Grundlagen. Marburg.
Cassirer, E. (1923a). *Function and Substance and Einstein Theory of Relativity* (W. C. Swaby & M. C. Swaby, Trans.). The Open Court Pub.
Cassirer, E. (1944). The Concept of Group in the Theory of Perception. *Philosophy and Phenomenological Research, 5*(1), 1–35.
Cassirer, E. (1953). *An Essay on Man.* Doubleday and.
Cassirer, E. (1965). *The Philosophy of the Enlightenment.* Beacon Press.
Cassirer, E. (2012). Form and Technology. In A. S. Hoel & I. Folkvord (Eds.), *Ernst Cassirer on Form and Technology.* Palgrave Macmillan.
Cassirer, E. (1923b-1932). *The Philosophy of Symbolic Forms.* Routledge.
Cieszkowski A. von. (1838). *Prolegomena zur Historiosophia.* Berlin.
Cohen, G. A. (1978). *Karl Marx's Theory of History — A Defense.* Princeton University Press.
Cole, H. S. (Ed.). (1973). *Models of Doom. A Critique of the Limits to Growth.* Universe Books.
Copleston, S. J. F. (1963-1965). *A History of Philosophy: Modern Philosophy.* Image Books.
Descartes. (1978). Discourse on Method. Principles of Philosophy In *The Philosophical Works of Descartes* (Vol. 1, E. S. Haldane & T. R. G. Ross, Trans.). Cambridge University Press.

Descartes. (1985). Arguments Demonstrating the Existence of God and the Distinction Between Soul and Body, Drawn Up in Geometrical Fashion. In J. Cottingham, R. Stoothoff, & D. Murdoch (Eds.), *The Philosophical Writings of Descartes* (Vol. 3). Cambridge University Press.

Descartes. (1988). Principles of Philosophy and Discourse on Method. In Descartes *Selected Philosophical Writings* (J. Cottingham, R. Stoothoff, & D. Murdoch, Eds.). Cambridge University Press.

Descartes. *Qeuvres de Descartes publees par Charles Adam et Paul Tannery* (Vol. 1).

DeGrood, D. (1980). *Marxism, Science and the Movement of History*. Gruner.

Deutsch-Französische Jahrbücher. (1844). https://www.marxists.org/archive/marx/works/1843/letters/43_09-alt.htm

Dodds, E. R. (1951). *The Greeks and the Irrational*. University of California Press.

Ihde, D. (1990). *Technology and the Lifeworld*. Indiana University Press.

Ihde, D. (1991). *Instrumental Realism*. Indiana University Press.

Dostoevsky, F. M. (1974). Бесы [Demons] In F. M. Dostoevsky (Ed.), Полное Собрание Сочинений [Collected Works] (Vol. 10). Изд. Наука [Science Pub.].

Dostoevsky, F. M. (1976). Братья Карамазовы *The Brothers Karamazov* In F. M. Dostoevsky (Ed.) Полное Собрание Сочинений [Collected Works] (Vol. 14). Изд. Наука [Science Pub.].

Dreyfus, H. L. (1995). Heidegger on Gaining a Free Relation to Technology. In A. Feenberg & A. Hannay (Eds.), *Technology and the Politics of Knowledge*. Indiana University Press.

Du Bois-Reymond, E. (1874, May). The Limits of Our Knowledge of Nature. *Popular Science Monthly*, 5.

Dupre, L. (1966). *The Philosophical Foundations of Marxism*. Harcourt, Brace and World.

Engels, F. (1954). *Dialectics of Nature*. Foreign Languages Publishing House.

Engels, F. (1972). Preliminary Theses on the Reform of Philosophy In *The Fiery Brook: Selected Writings of Ludwig Feuerbach* (Z. Hanfi, Trans.). Anchor Books.

Engels, F. (1976). "Principles of Communism," "Draft of a Communist Confession of Faith". In K. Marx & F. Engels (Eds.), *Collected Works* (Vol. 6). International Publishers.

Engels, F. (1987). Anti-Dühring. In K. Marx & F. Engels (Eds.), *Collected Works* (Vol. 25). International Publishers.

Engels, F. (1990). Ludwig Feuerbach and the End of Classical German Philosophy (1886). In K. Marx & F. Engels (Eds.), *Collected Works* (Vol. 26). International Publishers.

Eliade, M. (1957). *Das Heilige und das Profane*. Rowohlts Taschenbuchverlag.

Feenberg, A. (2002). *Transcending Technology* (2nd ed.). Oxford University Press.

Feenberg, A. (N.D.) *From Essentialism to Constructivism: Philosophy of Technology at the Crossroads*. https://web.archive.org/web/20061205020238/http://www-rohan.sdsu.edu/faculty/feenberg/talk4.html

Feuerbach, L. (1959-1960). *Sämtliche Werke* (12 Vols.) (2nd. ed, W. Bolin & F. Jodl, Eds.). Stuttgart.
Feuerbach, L. (1957). *The Essence of Christianity* (M. Evans, Trans.). Harper and Row Publishers.
Feuerbach, L. (1972). Preliminary Theses on the Reform of Philosophy. In *The Fiery Brook: Selected Writings of Ludwig Feuerbach*. Anchor Books.
Feuerbach, L. (1843). The Principles of the Philosophy of the Future. In *The Fiery Brook: Selected Writings of Ludwig Feuerbach* (Z. Hanfi, Trans.). Anchor Books.
Feynman, R. (1994). *The Character of Physical Law*. Modern Library.
Fichte, J. G. (1794). Einige Vorlesungen über die Bestimmung des Gelehrten. In I. G. Fichte (1845-1846) *Sammtlihe Werke* (Vol. 6, I. H. Fichte (junior), Ed.).
Fichte, J. G. (1845-1846). *Sammtliche Werke* (Vol. 2, 4, 6, J. G. Fichte (junior), Ed.).
Fichte, J. G. (1910). *The Vocation of Man*. Open Court Pub.
Fichte, J. G. (1977). *Characteristics of the Present Age* (Daniel N. Robinson, Ed.). University Publications of America.
Fichte, J. G. (1982). *The Science of Knowledge* (P. Heath & J. Lachs, Eds., Trans.). Cambridge University Press.
Fichte, J. G. (N.D.) Fichte, J. G. *Briefwechsel* (Bd. 2, hrsg. Von Hans Schulz).
Forrester, J. (1971). *World Dynamics*. Wright-Allen Press.
Fried, A., & Sanders, R. (Eds.). (1993). *Socialist Thought: A Documentary History*. Columbia University Press.
Fromm, E. (1968). *The Revolution of Hope—Toward a Humanized Technology*.
Fromm, E. (1971). *Marx's Concept of Man: With a Translation from Marx's Economic and Philosophical Manuscripts* (T. B. Bottomore, Trans.) (17th print.). Frederick Ungar Pub.
Fukuyama, F. (1992). *The End of History and the Last Man*. Free Press.
Fukuyama, F. (2007, July, 2). The End of History Revisited. *Seminars About Long-Term Thinking*. https://www.youtube.com/watch?v=w240nD5whsE
Fukuyama, F. (2017, April, 17). The End of International Liberal Order? *Taipei*. https://www.youtube.com/watch?v=scAzukYHJjY&list=PLsosaXrD-IswaVI YiygI6ZqYWnlKwin5P&index=45
Гайденко, П. П. (1979). Философия Фихте и Современность [Gaidenko, P. P. *Fichte's Philosophy and Modernity*]. Издательство Мысль.
Gebhardt, J. (1962). Karl Marx und Bruno Bauer. In *Politische Ordnung und menschliche Existenz*. Munich.
Gehlen, A. (1961). *Anthropologische Forschung*. Reibek.
Gehlen, A. (1980). *Man in the Age of Technology*. Columbia University Press.
Gehlen, A. (1977). *Technology and the Human Condition*. St. Martin's Press.
Giovanni Pico della Mirandola. (1956). *Oration on the Dignity of Man*. Henry Regnery.

Gorelic, G. (1983, July). Bogdanov's Tektology: Its Nature, Development, and Influence. *Studies in Soviet Thought*, 39–58.
Gorelic, G. Bogdanov's 'Tektology', General System Theory and Cybernetics. *Cybernetics and Systems: An International Journal, 18*(2), 157–185.
Gould, C. (1978). *Marx's Social Ontology—Individuality and Community in Marx's Theory of Social Reality*. MIT Press.
Habermas, J. (1971). *Knowledge and Human Interests* (J. J. Shapiro, Trans.). Beacon Press.
Haym, R. (2011)[1857]. *Hegel Und Seine Zeit*. Nabu Press.
Hegel, G. W. F. (1927). Philosophische Propadeutik In G. W. F. Hegel (Ed.), *Samtliche Werke* (Bd. 3, Hrsg. Von H. Glockner).
Hegel, G. W. F. *Science of Logic* (Vol. 2, pp. 413–414, W. H. Johnston & L. G. Struthers, Trans.). The Macmillan Co.
Hegel, G. W. F. (1942). *Philosophy of Right* (T. M. Knox. Trans.). Clarendon Press.
Hegel, G. W. F. (1962). *Lectures on the Philosophy of Religion* (Vol. 1-3). Routledge and Kegan Paul.
Hegel, G. W. F. (1963). *Lectures on the History of Philosophy* (Vol. 3, E. S. Haldane & F. H. Simson, Trans.). Humanities Press.
Hegel, G. W. F. (2010). *The Science of Logic*. Cambridge University Press.
Hegel, G. W. F. (1971). *The Phenomenology of Mind* (J. B. Baillie, Trans.). Humanities Press.
Hegel, G. W. F. (1802–1803). On the Scientific Ways of Treating Natural Law, on its Place in Practical Philosophy, and its Relation to the Positive Sciences of Right In G. W. F. Hegel (Ed.), *Political Writings* (L. Dickey & H. B. Nisbet, Eds.). Cambridge University Press.
Hegel, G. W. F. (2018). *Phenomenology of Spirit*. Cambridge University Press.
Heidegger, M. (1954). Die Frage nach der Technik. In *Die Künste im Technischen Zeitalter*. München.
Heidegger, M. (1962). *Being and Time*. Harper and Row.
Hess, M. (1961). *Philosophische und sozialistische Aufsatze, 1837-1850* (Cornu, Ed.). Berlin.
Hess, M. (1993). The Philosophy of the Act. In A. Fried & R. Sanders (Eds.), *Socialist Thought: A Documentary History*. Columbia University Press.
Hess, M. (2004). Consequences of the Revolution of the Proletariat. In M. Hess (Ed.), *The Holy History of Mankind and Other Writings* (S. Avineri, Ed.). Cambridge University Press.
Hilbert, D. (1926). On the Infinite. *Mathematischen Annalen, 95*, 161–190.
Hilbert, D. (1935). *Gesammelte Abhandlungen* (Vol. 3). Verlag Springer.
(Rabbi) Hillel. *Mishna: Pirkei Avot*, 13.
Horgan, J. (1996). *The End of Science*. Broadway Books.
Hook, S. (1943). *The Hero in History*. The John Jay.
Hook, S. (1936). *From Hegel to Marx*. The Humanities Press.

Huntington, S. P. (2011). *The Clash of Civilizations*. Simon & Schuster.
Husserl, E. (1970). *The Crisis of European Sciences and Transcendental Phenomenology* (D. Carr, Trans). Northwestern University Press.
Husserl, E. (1980). *Collected Works*. (Vol. 1). *Phenomenology and the Foundation of Sciences* (3rd book). *Ideas Pertaining to a Pure Phenomenology and to a Phenomenological Philosophy*. Martinus Nijhoff Publishes.
Ihde, D. (1978). *Technics and Praxis. A Philosophy of Technology*. D. Reidel Publishing Company.
Kant, I. (1929). *Critique of Pure Reason* (N. K. Smith, Trans.). Macmillan.
Kant, I. (1963). Idea for a Universal History from a Cosmopolitan Point of View. In I. Kant (Ed.), *On History*. The Bobbs-Merrill Co.
Kant, I. (1966). *Prolegomena to Any Future Metaphysics* (G. Lucas, Trans.). Manchester University Press.
Kant, I. (1979). *The One Possible Basis for a Demonstration of the Existence of God*. Abaris Books.
Kant, I. Beantwortung der Frage: Was ist Aufklanrung? In I. Kant (Ed.), *Werke* (Vol. 4, E. Cassirer, Ed.).
Kant, I. (2012). *The Critique of Practical Reason*. Dover Pub.
Kapp, E. (1877). *Gründlinien einer Philosophie der Technik--Zur Entstehungsgeschichte der Cultur aus neuen Gesichtspunkten*, Braunschweig; translated as Kapp, E. (2018). *Elements of a Philosophy of Technology: On the Evolutionary History of Culture* (3rd ed.). University of Minnesota Press.
Kats, Y. (2004). Bogdanov and the Limits to Growth Debate. *European Legacy*, 9(3), 305–317.
Katz, C. (1994, Summer). The Greek Matrix of Marx's Critique of Political Economy. *History of Political Thought.*, 15(2), 229–248.
Kitchin, R. (2014, April). Big Data, New Epistemologies and Paradigm Shift. *Big Data & Society*, 1(1), 1–12. https://doi.org/10.1177/2053951714528481
Kolakowski, L. (1978). *Main Currents of Marxism* (3 Vols.) (P. S. Falla, Trans.). Clarendon Press.
Koyrè, A. (1971). *Descartes und die Scholastic (1923)*. Bouvier Verlag H. Grundmann.
Latour, B. (1987). *Science in Action*. Harvard University Press.
Leibniz, G. W. (1697). *Animadvesiones in Partem Generalem Principiorum Cartesianorum*.
Leibniz, G. W. (1898). Monadology. In *Leibniz: The Monadology and Other Philosophical Writings* (R. Latta, Trans.). Oxford University Press.
Leibniz, G. W. (1996). *New Essays on Human Understanding*. Cambridge University Press.
Leibniz, G. W. (N.D.). *Principes de la Nature et de la Grace Fondees en Raison* in *Die Philosophischen Schriften Von Gottfried Wilhelm Leibniz* (1875–1890) (Bd. VI, pp. 598–606).

Liebich, A. (1979). *Between Ideology and Utopia: The Politics and Philosophy of August Cieszkowski.* D. Reidel.
Leiss, W. (1972). *The Domination of Nature.* Braziller.
Lenin, V. I. (1920). *Materialism and Empirio-Criticism.* Foreign Languages Publishing House.
Lenin, V. I. (1941-1962). *Sochineniya* [Works] (4th ed., 40 Vols.). Progress Pub.
Lenin, V. I. (1963). Filosofskie Tetradi (Philosophical Notebooks) In V. I. Lenin (Ed.), *Polnoye Sobranie Sochineniy* [Collected Works]. Gos. Izd. Polit. Lit.
Lenin, V. I. (1958-1965). *Polnoye Sobraniye Sochineniy* (5th ed., 55 Vols.). Gos. Izd. Polit. Lit.
Lenin, V. I. (1963-1964). *Selected Works* (3 Vols.). Progress Pub.
Lenin, V. I. (1970). *Collected Works* (45 Vols.). Progress Pub.
Lenin, V. I. (1975). Three Sources and Three Component Parts of Marxism. In C. Tucker (Ed.), *The Lenin Anthology.* Norton & Co.
Livergood, N. (1967). *Activity in Marx's Philosophy.* N. Nijhoff.
Löwith, K. (Ed.). (1962). *Die Hegelsche Linke.* Frommann.
Löwith, K. (1988). *Von Hegel zu Nietzsche.* J. B. Metzler.
Lobkowicz, N. (1967a). *Theory and Practice: History of a Concept from Aristotle to Marx.* Notre Dame University Press.
Lobkowicz, N. (Ed.). (1967b). *Marx and the Western World.* Notre Dame University Press.
Lukacs, G. (1971). *History and Class-Consciousness* (R. Livingstone, Trans.). MIT Press.
Lukacs, G. (1986). *Prolegomena Zur Ontologie des Gesellschaftlichen Seins.* Luchterhand.
Mach, E. (1991)[1906]. *Die Analyse der Empfindungen.* Wissenschaftliche Buchgesellschaft.
Maimon, S. (2010). *Essay on Transcendental Philosophy.* Continuum.
Maimonides. (N.D.). *The Guide for Perplexed.* (4th ed, M. Friedlander, Trans.). E. P. Dutton & Co.
Malthus, T. (2015)[1798]. *An essay on the Principle of Population.* Penguin.
Mann, T. (2005). *Joseph and His Brothers.* Alfred A. Knopf.
Marcus, G. (1978). *Marxism and Anthropology.* Van Gorcum Assen.
Marcuse, H. (1941). *Reason and Revolution: Hegel and the Rise of Social Theory.* Oxford University Press.
Marcuse, H. (1969). The Foundations of Historical Materialism. In H. Marcuse (Ed.), *Studies in Critical Philosophy.* Beacon Press.
Marx, K. (1961). Economic and Philosophical Manuscripts. In E. Fromm (Ed.), *Marx's Concept of Man with a Translation from Marx's 'Economic and Philosophical Manuscripts'* (Bottomore, Trans.). Frederick Ungar.
Marx, K. (1973). *Grundrisse* (Nicolaus, Trans.). Vintage Books.

Marx, K. (1975a)[1844]. A Contribution to the Critique of Hegel's Philosophy of Right In K. Marx (Ed.), *Early Writings* (R. Livingstone & G. Benton, Trans.). Vintage Books.
Marx, K. (1975b). Economic and Philosophical Manuscripts In K. Marx. (Ed.), *Early Writings* (R. Livingstone & G. Benton, Trans.). Vintage Books.
Marx, K. (1978). The German Ideology, "On the Jewish Question," "Theses on Feuerbach," Preface to Critique of Political Economy. In D. McLellan (Ed.), *Karl Marx. Selected Writings.* Oxford University Press.
Marx, K. (1984). *Capital* (Vol. 3, F. Engels, Ed.). International Publishers.
Marx, K. (2010). Economic and Philosophical Manuscripts of 1844. In K. Marx & F. Engels (Eds.), *Collected Works.* Lawrence & Wishart.
Marx-Engels. (1927-1935). *Gesamtausgabe* (MEGA) (12 Vols., Ryazanow, Ed.). Berlin.
Marx-Engels. (1964-1968). *Werke* (39 Vols.). Berlin.
Marx-Engels. (1965-1966). *Sochineniya* (39 Vols.). Politizdat.
Marx-Engels. (1975-1995). *Collected Works* (47 Vols.). International Publishers.
Мартов, Ю. О. (1923). Мировой Большевизм [Martov. *The World Bolshevism*]. Iskra.
Mattessich, R. (1978). *Instrumental Reasoning and Systems Methodology.* D. Reidel.
Matustic, M. J. (1995). Kierkegaard's Radical Existential Praxis, or: Why the Individual Defies Liberal, Communitarian, and Postmodern Categories. In M. J. Matustic (Ed.), *Post-Modernity.* Indiana University Press.
McCarty, D. C. (2013). *David Hilbert and Paul du Boise-Reymond: Limits and Ideals.* oai: CiteSeerX.psu:10.1.1.366.4505.
McCutcheon, R. (1979). *Limits of a Modern World: A Study of the Limits to Growth Debate.* Butterworths.
McLellan, D. (1969). *The Young Hegelians and Karl Marx.* Macmillan.
Meadows, D. H., Meadows, D. L., et al. (1972). *The Limits to Growth.* Universe Books.
Meadows, D. H., Randers, J., & Meadows, D. L. (2004). *Limits to Growth: The 30-Year Update.* Chelsea Green Pub Co.
Medawar, P. B. (1984). *The Limits of Science.* Harper & Raw Publishers.
Memories and Proceedings of the "Manchester Literary and Philosophical Society. (1949-1950). Vol. 91.
Meyer, A. G. (1962). *Leninism.* Praeger.
Meyer, A. G. (1970). *Marxism: The Unity of Theory and Practice.* Harvard University Press.
Mill, J. S. (2015). *On Liberty* (2nd ed.). Oxford University Press.
Mills, C. W. (1956). *The Power Elite.* Oxford University Press.
Mises, L. (1985). *Liberalism in Classical Tradition* (3rd ed.). The Foundation for Economic Education & Cobden Press. (online 2002 ed. by Mises.org)
Moggach, D. (1994, July). Marx and German Idealism: Labor and the Transcendental Synthesis. *History of European Ideas., 19*(1–3), 137–143.

Von Neumann, J. (1966). *Theory of Self-Reproducing Automata*. University of Illinois Press.
Ninnes, L. E. (1993). Hegel, Marx and Idealistic Vapourizing. *Dialogue Hum*, 3(4), 97–111.
Noire, L. (1880). *Das Werkzeug und seine Bedeutung für die Entwicklungsgeschichte der Menschheit* (1968 reprint is available).
Novack, G. (1975). *Pragmatism vs Marxism*. Pathfinder Press.
O'Brien, R. C. (1995). Marxist Eschatology and Christian Eschatology: Marx and the Story of Man's Deliverance. In R. P. Francis, Lang (Ed.), *Christian Humanism: International Perspectives*.
Odajnyk, W. (1965). *Marxism and Existentialism*. Doubleday-Anchor Press.
Oizerman. (1962). *Formirovanie Filosofii Marksizma*. Izdatel'stvo Sotzial'no Ekonomicheskaya Literatura.
Plekhanov, G. V. (1906). *Kritika Nashich Kritikov* (Critique of Our Critics). Obshchestvennaya Polza.
Plekhanov, G. V. (1922-1927). *Sochineniya* [Works] (24 Vols., D. Ryazanov, Ed.). Moscow and Leningrad.
Plekhanov, G. V. (1934). *The Essays on the History of Materialism*. (R. Fox & J. Lane, Trans.). The Bodley Head.
Plekhanov, G. V. (1956-1958). *Izbranniye Filosofskie Proizvedeniya* (5 Vols.). Gosudarstvennoe Izdatel'stvo Politicheskoi Literatutri [*Selected Philosophical Works*, 5 Vols.].
Peccei, A. (1977). *The Human Quality*. Pergamon Press.
Plato. (1934). *The Laws of Plato* (Book XII, 942, A. E. Taylor, Trans.). J. M. Dent & Sons Ltd.
Poincare, H. (2018). *The Value of Science*. Forgotten Books.
Popper, K. R. (1965). Prediction and Prophecy in Social Sciences. In K. R. Popper (Ed.), *Conjectures and Refutations: The Growth of Scientific Knowledge*. Harper & Row.
Popper, K. R. (1966). *The Open Society and its Enemies* (5th ed.). Princeton University Press.
Popper, K. R. (1994). Epistemology and Industrialization. In K. R. Popper (Ed.), *The Myth of the Framework*. Routledge.
Prigogine, I., & George, C. (1983). The Second Law as a Selection Principle: The Microscopic Theory of Dissipative Processes in Quantum Systems. *Proceedings of the National Academy of Sciences of the Unites States of America, 80*, 4590–4594.
Prigogine, I., & Stengers, I. (1984). *Order out of Chaos*. Heinemann.
Proclus. (1992). *The Elements of Theology* (2nd ed.). Clarendon Press.
Putnam, H. (1983). Analyticity and APriority: Beyond Wittgenstein and Quine. In *Realism and Reason. Philosophical Papers* (Vol. 3). Cambridge University Press.

Radrizzani, I. (1996). Fichte's Transcendental Philosophy and Political Praxis. In T. Rockmore (Ed.), *New Perspectives on Fichte*. Humanities Press.
Rapp, F. (1981). *Analytical Philosophy of Technology*. D. Reidel.
Rawls. (1999). *A Theory of Justice*. Belknap Press.
Reichenbach, H. (1991). *The Direction of Time*. University of California Press.
Rescher, N. (1999). *The Limits of Science*. University of Pittsburgh Press.
Ridker, R. G. (1973, December). To Grow or not to Grow. *Science, 182*(4119), 1315–1318.
Rockmore, T. (1963, March). Merleau-Ponty, Marx, and Marxism: The Problem of History. *Studies in Eastern European Thought., 48*(1), 63–81.
Rockmore, T. (1978). Marxian Praxis. *Philosophy and Social Criticism*, 5.
Rockmore, T. (1980). *Fichte, Marx and the German Philosophical Tradition*. Southern Illinois University Press.
Ruge, A. (1847). *Sämtliche Werke*. Mannheim.
Saito, K. (2023). *Marx in the Anthropocene*. Cambridge University Press.
Sartre, J. P. (1963). *The Problems of Method*. Methuen & Co.
Schelling, F. W. J. (1975). *Zur Geschichte Der Noueren Philosophie*. Wissenschaftliche Buchgesellschaft.
Schelling, F. W. J. (1978). *System of Transcendental Idealism* (P. Heath, Trans.). University Press of Virginia.
Schelling, F. W. J. (1980). *Philosophical Letters on Dogmatism and Criticism* (Letter 4). In F. W. J. Schelling (Ed.), *The Unconditional in Human Knowledge: Four Early Essays, 1794–1796* (F. Marti, Trans.). Bucknell University Press.
Schelling, F. W. J. (1989). *The Philosophy of Art*. University of Minnesota Press.
Schelling, F. W. J. (2006). *Philosophical Investigations into the Essence of Human Freedom and the Related Matters* (J. Love & J. Schmidt, Trans.). State University of New York Press.
Schelling, F. W. J. (2018). *Einleitung in die Philosophie der Mythologie*. Forgotten Books.
Schmidt, A. (1971). *The Concept of Nature in Marx* (Fowkes, Trans.). B. NBL.
Schroyer, T. (1973). *The Critique of Domination*. G. Braziller.
Schouls A. P. (1991). Descartes and the Idea of Progress. In *Rene Descartes: Critical Assessments* (Vol. 1, J. D. Moyal, Trans.). Routledge.
Shub, D. (1966). *Lenin. A Biography*. Penguin.
Schumacher, E. F. (1999). *Small is Beautiful: Economics as if People Mattered*. Hartley & Marks.
Simon, J. L. (1996). *The State of Humanity*. Wiley-Blackwell.
Sochor, Z. A. (1981, February). Was Bogdanov Russia's Answer to Gramsci? *Studies in Soviet Thought, 22*, 59–81.
Spencer, H. (1981). The Social Organism, Specialized Administration. In Herbert Spencer. *The Man vs the State*. Liberty Classics.
Spengler, O. (2013). *Prussianism and Socialism*. Isha Books.

Spengler, O. (1976). *Man and Technics: A Contribution to a Philosophy of Life*. Greenwood Press. (Reprint of 1932 ed. by Alfred A. Knopf, Inc.).
Spiniza, B. (1927). *Ethics*. (W. H. White & A. H. Stirling, Trans.). Oxford University Press.
Spiniza, B. (2001). *Theologico-Political Treatise* (S. Shirley, Trans.). Hackett Pub.
Spiniza, B. (N.D.). On the Improvement of the Understanding. In B. Spinoza (Ed.), *Improvement of the Understanding, Ethics and Correspondence* (R. H. M. Elwes, Trans). Beling Tetens Publisher.
Stent, G. S. (1969). *The Coming of the Golden Age*. Natural History Press.
Stepelevich, L. S. (Ed.). (1983). *The Young Hegelians: An Anthology*. Cambridge University Press.
Sterling, J. H. (2017). *The Secret of Hegel*. Forgotten Books (Reprint of 1898 2nd ed.).
Stirner, M. (1912)[1845]. *The Ego and His Own* (Der Einzige und sein Eigentum) (S. T. Byngton, Trans.). London.
Strauss, D. F. (1841). *Streitschriften* (Bd. 3). Tubingen.
Strauss, D. F. (1860). The Life of Jesus Critically Examined (Vol. 2, M. Evans, Trans.). Calvin Blanchard.
Takhtadjan, A. L. (1972). *Tektologiya: Istoriya i Problemy* [Tektology: History and Problems]. *Systemnye Issledovaniya* [Systems Research Yearbook]. Isdatel'stvo Nauka.
Talmon, J. L. (1960). *Political Messianism: The Romantic Phase*. Secker & Warburg.
Thomson, E. (1994). The Sparks That Dazzle Rather Than Illuminate: A New Look at Marx's 'Theses on Feuerbach'. *Nature of Social Thought*, 7(3), 299–323.
Tillich, P. (1951). *Systematic Theology*. University of Chicago Press.
Toews, J. E. (1980). *Hegelianism: The Path Toward Dialectical Humanism, 1805–1841*. Cambridge University Press.
Torrance, J. (1995). *Karl Marx's Theory of Ideas*. Cambridge University Press.
Toynbee, A. J. (1948). *Civilization on Trial*. Oxford University Press.
Trotsky, L. (1925). *Lenin*. Minton, Balch & Co.
Трубецкой, С. Е. (1991). *Минувшее* (The Past). ДЭМ.
Tucker, R. C. (1961). *Philosophy and Myth in Karl Marx*. Cambridge University Press.
Tucker, R. C. (1970). *The Marxian Revolutionary Idea*. Norton.
Vysheslavtzev, B. P. (1994). *Etika Preobrazennogo Erosa* (Ethics of Transformed Eros). Izdatelstvo Respublika.
Vysheslavtzev, B. P. (1995). *Filosofskaya Nischeta Marksizma* [The Philosophical Poverty of Marxism] In B. P. Vysheslavtzev (Ed.) *Sochineniy*a [Works]. Raritet.
Wallerstein, I. (2004). *World-Systems Analysis: An Introduction*. Duke University Press.
Wartofsky, M. (1967). Comments on Irving Fetscher and Gajo Petrovic on the Young and the Old Marx and on Alienation. In N. Lobkowicz (Ed.), *Marx and the Western World*. Notre Dame University Press.

Wartofsky, M. (1975). Art as Humanizing Praxis. *Praxis—A Radical Journal of the Arts, 1*.
Wartofsky, M. (1977). *Feuerbach*. Cambridge University Press.
Waser, R. (1994). *Autonomie des Selbstbewusstseins: Eine Untersuchung zum Verhältnis von Bruno Bauer und Karl Marx (1835–1843)*. Francke.
Wendling, A. E. (2009). *Karl Marx on Technology and Alienation*. Palgrave Macmillan.
Weyl, H. (1949). *Philosophy of Mathematics and Natural Science* (O. Helmer, Trans.). Princeton University Press.
Weyl, H. (2009). *Mind and Nature: Selected Writings on Philosophy, Mathematics and Physics* (P. Pesic. Ed.). Princeton University Press.
Weyl, H. (2012). Levels of Infinity. In H. Weyl (Ed.), *Levels of Infinity*. Dover Pub.
Whitehead, A. N. (1963). *Science and the Modern World*. New American Library.
Wiener, N. (2013). *Cybernetics* (2nd ed.). Martino Fine Books.
Winner, L. (1986). *The Whale and the Reactor*. University of Chicago Press.
Wright, E. (1986, Summer). Dialectical Perception: A Synthesis of Lenin and Bogdanov. *Radical Philosophy*, 9–16.
Wetter, G. (1966). *Dialectical Materialism* (P. Heath, Trans.). Praeger.
Yassour, A. (1981, February). Lenin and Bogdanov. *Studies in Soviet Thought, 22*, 1–32.
Yassour, A. (1983, July). The Empiriomonist Critique of Dialectical Materialism: Bogdanov, Plekhanov, Lenin. *Studies in Soviet Thought, 26*, 21–38.
Yassour, A. (1984, April). Bogdanov-Malinovsky on Party and Revolution. *Studies in Soviet Thought, 27*, 225–236.
Yassour, A. (1986, April). Philosophy—Religion—Politics: Borochov, Bogdanov, Lunacharsky. *Studies in Soviet Thought, 31*, 199–230.
Yovchuk, M. T. (1960). *Plekhanov i Ego Trudy po Istorii Filosofii* [Plekhanov and His Works in the History of Philosophy]. Moscow.
Zavadskiy, K. M., & Kolchinskiy, E. I. (1977). *Evolutziya Evolutzii* [Evolution of the Evolution]. Nauka.
Zenkovsky, V. V. (1953). *A History of Russian Philosophy*. Routledge and Kegan Paul.
Zlociti, T. (1921). *Moses Hess, Vorkämpfer des Sozialismus und des Zionismus* (2nd ed.). L. Lamm.

Name Index[1]

A
Ackermann, W., 137n68
Adler, Max, 72, 106, 107
Anderson, C., 132n54
Anselm, St., 23
Aristotle, 19, 91n31
Arthur, C. J., 91
Auger, Pierre, 127, 128, 134
Augustine, St., 24

B
Bachelard, G., 135, 137, 138
Bacon, Fransis, 27
Barrow, J. D., 129n44
Bauer, Bruno, x, 9–11, 47, 48, 51, 58–64, 61n24, 63n29, 68, 70
Bazhanov, B. G., 97n44
Bergson, Henry, 144
Bernstein, Edward, 13, 35, 60, 98, 102
Bertalanffy, Ludwig, 116
Bogdanov, A., xi, 81, 84n14, 87, 98, 108–111, 108n71, 114–121, 120n19, 124, 130, 134
Bohr, Niels, 33, 135n64
Born, Max, 135n64
Brouwer, L. E. J., 32n31, 38n41, 129n45, 139, 139n76
Brzozowski, Stanislaw, 108
Bury, J. B., 4n9, 9n23

C
Calude, C. S., 132
Cassirer, Ernst, 31n30, 45n56, 114, 135n63, 137, 137n70
Cieszkowski, A. von, 68
Cole, H. S., 123n28
Copleston, F. S. J., 6n15, 28n20, 29n24

[1] Note: Page numbers followed by 'n' refer to notes.

162 NAME INDEX

D
De Broglie, L., 138n71
Descartes, Rene, ix, 3, 20, 22–27, 22n8, 25n13, 28n21, 30, 30n28, 31, 52, 121, 136, 137
Djilas, M., 101n53
Dodds, E. R., 97n45
Dostoevsky, F. M., 18, 24–25n13
Dreyfus, H. L., 141, 141n81
Du Boise-Reymond, Paul, 129n45

E
Einstein, Albert, 108, 108n70, 110, 113, 114n1, 120n19, 133, 138
Engels, Friedrich, 10, 12, 47n59, 51, 59, 62, 63, 70, 71, 73n61, 75, 85, 89, 89n25, 92n37, 94, 102–104, 104n62, 106, 107, 121

F
Feenberg, A., 141, 141n81, 142n82
Feuerbach, Ludwig, x, 9, 11, 48, 50–52, 62–69, 71–75, 73n61, 74n62, 81, 82, 87, 103, 105–107
Feynman, Richard, 17
Fichte, J. G., 6–8, 6n16, 8n21, 10, 11, 20, 21, 28n21, 32–34, 36, 38–43, 38n42, 46n58, 49–51, 53–55, 59, 60, 63, 91, 102, 109, 122
Fukuyama, Francis, vii

G
Gaidenko. P. P., 6n15
Galen, A., 87
Gendron, B., 15n43, 18n48
Godel, Kurt, 17, 115, 128, 129

H
Habermas, Jurgen, 91, 92, 142n82
Hegel, G. W. F., ix, x, 6n15, 7–12, 8n21, 11n26, 21, 21n5, 26, 28n21, 32, 35, 40–45, 44n52, 47–51, 53–55, 57, 59–61, 64, 66, 68, 72, 74–76, 81, 83, 84, 93, 99, 101, 103, 105, 116, 140
Heidegger, Martin, 18, 86, 114, 127, 140, 142n82, 145n88
Heisenberg, Werner, 33, 110, 135n64, 137
Hess, Moses, x, 48, 51, 61–63, 68–71
Heyting, A., 139n75
Hilbert, David, 129n45, 132, 133, 133n56, 139n76
Hillel, Rabbi, 145n89
Hook, Sidney, 63n29, 70
Horgan, John, 125, 126
Hume, David, 4, 25, 29, 29n24, 29n25, 31, 31n30, 107, 108, 132
Huntington, S. P., 15
Husserl, Edmund, 136

I
Ihde, Don, 58n18, 125, 125n34

K
Kant, Immanuel, viii, xii, 4–8, 6n15, 20, 26, 28–31, 28n21, 29n25, 31n30, 32n31, 33–36, 40, 41, 43, 44, 46, 49, 53–57, 59, 60, 67, 85, 105–108, 115, 129, 132, 135n63, 136, 137, 139, 141, 143
Kapp, Ernst, 32, 87, 88
Kautsky, Karl, 98, 99, 101, 102
Kleene, S. C., 132
Kojev, A., 44

NAME INDEX

Kolakowski, Leszek (Leslie), 52n5, 60, 92n37, 98, 103n59, 108n71
Koyre, A., 24, 136

L
Labriola Antonio, 108
Latour, Bruno, 125, 125n34
Lavrov, Petr, 94
Leibniz, G. W., 4, 16n16, 26–28, 28n20, 30n28, 31n30, 38, 40, 53, 67
Lenin, Vladimir, 2, 11n26, 84, 98–104, 103n59, 106–108, 108n71, 110, 128
Lorentz, Konrad, 128, 134
Lucacs, Gyorgy (George), 74n63

M
Mach, Ernst, xi, 107, 108, 110, 114, 117, 119, 135n63
Maimon, Solomon, 36n37
Malthus, Thomas, 121
Mann, Thomas, 58
Marcus, G., 91n35
Martov, Julius, 95, 100
Marx, Karl, vii–xii, 1n1, 3, 4, 7, 10–12, 10n25, 11n26, 21, 21n5, 22, 26–28, 31, 32, 34, 35, 41, 45–49, 47n59, 51, 58, 59, 61–64, 63n29, 67–77, 68n45, 69n47, 74n63, 79–111, 92n37, 104n62, 108n71, 109n75, 114, 116–121, 128, 134, 138, 141, 142, 144
McLellan, David, 52n5, 59, 60, 63, 70, 72
Meadows, D. H., 115n2
Mikhaylovsky, Nikolai, 94
Mill, John Stuart, 6, 109
Mills, C. W., 14

Mirandola, Giovanni Pico della, 19
Mises, Ludwig von, 8, 101

N
Neumann, John von, 84, 129
Nietzsche, Friedrich, 73, 127, 128
Noire, Ludwig, 32, 87, 88

P
Plank, Max, 136n66
Plato, 11n26, 19, 32, 97, 101, 116, 117
Plekhanov, G. V., 103, 103n59, 108n71
Poincare, Henry, 113, 114, 137
Popper, Karl, 11n26, 61n24
Prigogine, I., 118n14, 119n14
Proclus, 39n43
Putnam, Hilary, 14n37

R
Rapp, F., 143n84
Rawls, John, 99
Rescher, Nicholas, 129–132, 134
Ridker, R. G., 124
Rockmore, Tom, 22n8, 28n21
Rousseau, Jean-Jacques, 8, 8n21, 27, 68, 101n55
Ruge, Arnold, x, 51, 61–63

S
Saito, K., 58n18, 94, 95
Sartre, Jean-Paul, 93, 97
Schelling, F. W. J., 6n16, 7, 16, 29, 30, 34–36, 40–43, 45–47, 46n58, 57–60, 140
Schmidt, A., 67, 104
Schumacher, E. F., 143

Scott, Duns, 23
Simon, J. L., 122
Spencer, Herbert, 108, 111, 116–119, 128n42
Spengler, Ostwald, ix, 2, 8, 13, 15, 94
Spinoza, Baruch, 26, 27, 38, 41, 42, 50, 52, 53, 57, 103, 143
Stent, G. S., 126–128, 134
Stirner, Max, 9, 72–74, 73n61
Strauss, David, x, 46n58, 48, 50, 54–59, 64, 65n39

T
Tillich, Paul, 52n7
Toynbee, A. J., viii
Trotsky, Leon, 50n2, 97, 97n44
Trubetskoy, S. E., 97n44
Tucker, R. C., 12, 84

V
Visheslavtsev, B. P., 25n14

W
Wallerstein, I., 15
Wartofsky, Marx, 72, 92n37
Wendling, A. E., 76
Weyl, Hermann, 113, 114n1, 138, 139, 139n76
Wiener, Norbert, 116
Winner, Langdon, 16

Z
Zasulich, Vera, 94
Zenkovsky, V. V., 103n59, 106n67
Zeno, 132, 133n56

Subject Index[1]

A

Absolute, 7–12, 21, 24, 38, 39, 41–43, 45–47, 51, 53–55, 57, 60, 66, 77
  dialectic of, 12, 21, 40, 45, 54
  and Fichte, 6, 7, 38–40, 43, 45–47, 53–55, 92
  and Hegel, 7–10, 12, 21, 40–45, 51, 53–55, 59, 60, 66
  I, 6, 7, 21, 37, 40, 92
  and idealism, 38, 45
  as Neo-Platonic First Principle, 39
  schema of, 39, 43 (*see also* Fichte)
  as spirit, 7, 11, 21, 41, 43, 54, 66, 136
  See also God
Activity, 4, 6, 20, 26, 28, 31–37, 39, 41, 46, 63, 67, 69, 72, 74, 84, 89, 135, 137–139
  human, x, xi, 10, 21–24, 40, 45, 47, 51, 64, 75, 76, 85, 89, 92, 109, 110, 116, 119, 142
  infinite, 34, 92
  objective, 11, 21, 51, 75, 76, 82, 91, 92, 105, 107, 110
  universal, x, xi, 10, 11, 22, 47, 64, 76, 89
Actual infinity, *see* Infinity
Alienation, 64, 69–71, 69n51, 80, 93, 94, 97, 98, 109, 142
  and emancipation, 71
  vs. objectification, 69n51, 93, 97, 98
A priori, 25, 28n21, 29, 31, 32n31, 39, 93, 128, 129, 132, 139
  and evolution, 128
  forms of sensibility, 32, 33
  forms of understanding, 32, 33
  and Kant, 31, 129
  vs. a posteriori, 29, 32, 37, 76
  synthesis, 32, 33, 36, 139

---

[1] Note: Page numbers followed by 'n' refer to notes.

SUBJECT INDEX

Artificial Intelligence (AI)
   and Big Data, 131, 132, 141
   and data mining, 132, 141
   and machine learning, 131
   and operational complexity, 131, 132, 134

B

Being, x, 14, 17, 20, 22–26, 30, 32, 34, 37, 39, 57, 59, 60, 64, 66, 73, 81–84, 87, 88, 94–97, 107, 108, 114, 129, 133, 136, 139–141, 145
   consumer, 2, 94
   as Dasein, being-in-the-world, 29, 86, 114, 140
   and enframing, 86, 140
   for himself, 65, 66, 75, 81–83, 99, 105
   natural, 13, 67, 68, 71, 75, 76, 82, 86, 91, 104
   total, 11, 66, 67, 75, 80, 82, 85, 105
   universal, vii, viii, 2, 3, 10–13, 13n33, 65, 65n39, 72, 74–76, 83, 91, 92, 96, 99, 101, 104
   whole, 66, 68, 73
   See also Man

C

Capital, 16, 70, 94
   and capitalism, 90
   and capitalist society, 80
   and labor, 80
   as a universal form of property, 80
Cartesian, viii, ix, 3–5, 7, 25–28, 28n21, 32, 38, 42, 48, 52, 65, 77, 96, 111, 115, 124, 134, 138, 140
   circle, x, 20, 22, 24, 31, 48, 124, 136
   method, 22, 22n8
Circle, 51, 81–89
   Cartesian, x, 20, 22, 24, 30, 48, 124, 136
   epistemological, 22, 30, 124
   Fichtean, 21, 37
   Marxian, 48, 51, 81–89
Civil society, 5, 13, 15, 47
   and Hegel, 8, 9, 43, 70
   and Marx, 70, 71
   vs. social humanity, 70, 71
Club of Rome, 16, 114
   and limits to growth, 115, 123
   and systems approach, 114
   and technological progress, 17
   and zero growth, 16
Cogito, 7, 20, 22–24, 24n13, 28n21, 29, 31, 65–67
   Cartesian, 7, 22, 24n13, 29, 65
Communism, 12, 64, 69, 90, 103
   degrowth, 94
   scientific, viii, 7, 70, 74, 80, 81, 100
Consciousness, Self-consciousness, 39, 60, 62, 64–67, 75, 83, 88
   historical, 9–10, 46, 73
   mythological, myth-making, 46, 56–58
   social(ized), 82, 106
Construction, 40, 46, 92
   act of, 33, 36
   of concepts, 31
   and intellectual intuition, 31, 36
   and Kant, 4, 31–33, 36, 88, 139
   mathematical, 31, 31n30, 32, 32n31, 132, 139
Constructivism, 31n30, 140
   and Brouwer, 32n31, 139, 139n76
   and Weyl, 113, 114n1, 138, 139
Consumer, 18, 140, 141, 144
   being, 2, 13, 94
   society, 14, 15

Control, 142
   automatic, 117, 118 (*see also* Cybernetics)
   technical, 141, 142
   technological, 142
Conventionalism, 137
   Poincare, 113
   science, 31
Critique, 4, 7, 10, 15, 26, 29–31, 34, 40, 41, 46, 50, 53, 54, 57, 69, 71, 72, 74, 75, 99, 100, 105, 107, 113–145
   of reason, 28, 52
   social, ix, 9, 47, 48, 51, 59–68
   theological, x, 46n58, 48, 55, 59
Culture, 1, 18, 44, 144, 145
   of discipline, 5, 141, 143
   humane, 5, 143
   vs. nature, xi, 58
   of skill, 5, 141, 143
Cybernetics, xi, 114, 116, 177
   and Bogdanov, xi, 114
   and general theory of organization, xi, 114
   and Tektology, xi, 114

D
Dasein, 29
   and being, 140
   and enframing, 86, 140
   and Heidegger, 86, 114, 140
   and ontology, 140
   See also Being
Degrowth, *see* Growth
Democracy, 14
   and civil society, 8, 13
   liberal, 1, 1n1, 8, 13, 45, 100
Dialectic, 7, 79–111, 113–117, 130
   circular, 3, 21, 24, 51, 52, 81, 98
   close-ended, viii, 10, 28, 45, 47, 89, 118
   and dynamic equilibrium, 118
   Fichtean(ized), x, 9, 10, 21, 41, 43, 46–48, 59, 63, 81
   Hegelian, ix, x, 9, 10, 12, 22, 31, 37, 38, 47, 48, 51, 59–61, 75, 77, 81, 84, 84n14, 88, 91
   of history, 41–45, 47, 63, 118
   inverted, x, 11, 47, 51, 77
   of knowledge, 19–48
   of labor, ix, x, 3, 4, 11, 12, 21, 28, 35, 45, 47, 49–77, 81, 87–89, 97, 98, 118
   Marx's, ix, 4, 11, 12, 21, 28, 68, 81, 87, 88, 97, 104n62
   materialistic, x, 114
   and negation of the negation, 104, 118
   open-ended, 9, 51, 59–61
   self-referential, x, 12, 75, 81
   of technology, viii, 12, 47, 84, 85, 89
Dualism, 7, 29–35, 136
   Cartesian, viii, 5, 20, 25, 26, 52, 77, 96, 111, 115, 134, 138
   Kantian, 22, 33, 36, 39, 40, 56

E
Elites, 96, 101
   power, 14, 18, 97, 100, 101, 144
   ruling, 15
Emancipation, x, 15, 70, 71, 74, 80, 86, 93, 95–97, 142
   epistemological, 110
   social, 69, 110
Empiricism, 87, 110, 137
   and Hume, 107, 132
   and Kant, 29, 34
Empiriocriticism, 107
   and Bogdanov, xi, 109, 111, 114
   and Mach, 81, 108, 117
   See also Mach

End of history, vii, 1, 9, 10, 13–18, 44
  and end of science, 125, 132
  and Fukuyama, 1n1
End of science, ix, xi, 3, 17, 81, 111
  and end of history, 125, 132
  and Horgan, 125, 126
  and Rescher, 129–132, 134
  and Stent, 126–128, 134
  and technology, technological progress, 125, 127, 128, 130
  and technoscience, 124
Enlightenment, x, 2–4, 14, 21, 27, 40, 45, 49, 60, 73, 97, 114, 143
  and historical optimism, 3, 44, 113
  philosophic spirit of, 6
Entropy
  and Bogdanov, 119, 120
  and inexhaustible creativity, 119, 120, 124
  and second law of thermodynamics, 119, 120, 124
  and technological progress, 119, 124
Epistemology, xi, 22n8, 24–26, 40, 110, 124, 136
  and epistemological circle, 22, 30, 124
Essence, 9, 22, 44, 57, 62, 70, 80, 105, 109, 111, 128, 130, 135, 138–140, 143
  and existence, 43, 52, 58, 66, 73, 79
  and existentialism, 73
  and God, 52–54, 56, 61, 66, 69
  human, 51, 60, 64, 66, 68, 69, 71–74, 77, 82, 83, 85, 101
Existence, 2, 4, 6–10, 17, 20, 29, 31, 35, 38, 43, 46, 52, 56, 58, 63, 64, 72, 74, 83, 85, 87, 90–92, 121, 124, 140, 142
  actual *vs.* formal, 23
  contingent *vs.* necessary, 23, 68, 71

  and essence, 43, 52, 58, 66, 68, 73, 79
  and existentialism, 73
  of God, 9, 23, 24, 24–25n13, 29n25, 30, 38, 52, 62, 66
  and Heidegger, 114, 140
  and Stirner, 9, 73
Experience, viii, 20, 22, 26, 30, 33, 35, 37, 40, 46, 56, 93, 114, 116, 133, 135, 137, 144, 145
  completeness of, 105, 113
  complexes of, 108, 109, 119
  elements of, 108, 108n72, 109, 119, 136, 139
  mythological, 45, 58
  religious, 45, 58

F
Fetishism, 72
  cognitive, 109, 110
  of commodities, 70, 109
  epistemological emancipation, 110
  of matter, 109
  scientific, 109
Fichtean circle, 37
Freedom, 1, 2, 5–8, 10, 17, 18, 26, 34, 36–41, 43–45, 47, 49, 50, 53, 56, 60, 79, 81, 86, 96, 97, 101, 103, 136–145
  moral *vs.* material necessity, 20, 35
  philosophy of, 21, 34, 35, 47, 81
  (*see also* Fichte)
Free will, *see* Will

G
General systems theory, xi, 84n14, 114
General theory of organization, xi, 84n14, 111, 114, 116–118
  and the second law of thermodynamics, 111, 114

See also Bogdanov; Tektology
God, 2, 3, 5, 9, 17, 19–21, 24n13, 25, 26, 28–31, 38–40, 43, 50, 52–57, 61, 62, 64, 66, 69, 73, 103, 115
 and capital, 69
 ontological proof of, 23, 30 (*see also* Existence)
 as schema or knowledge, 39 (*see also* Fichte)
 as totality, 9, 66
 See also Absolute
Group, 10, 32, 48, 50, 64, 80, 98, 99, 114
 conventionalism, 137
 invariant, 137
 relations, 137
 See also Mathematics
Growth, vii, 12, 16, 79–111, 115, 122, 124
 constraints of, 16, 121
 and degrowth, 95
 economic, 15, 114, 121, 122
 exponential, 123
 and inexhaustible creativity, 118–121
 infinite, ix, xi, 2, 3, 13n33, 89–98, 110, 115, 118–121, 123
 limits of, ix, xi, 3, 16, 18, 81, 111, 121–124, 130
 of population, 121, 122
 technological, vii, 12, 16, 79–111, 115, 122, 124
 unbounded, 12–18, 122
 unlimited, 12, 89, 98, 122, 124
 zero growth, 16, 95, 115, 122, 123

H

History, xi, xii, 4–7, 11, 15, 17, 19–48, 54–60, 62, 63, 72, 76, 77, 80–82, 84–87, 89, 92, 101, 102, 104, 108, 121, 125, 136–145
 close-ended, x, 9, 21, 45, 47, 118
 end of, vii, 1, 9, 10, 13–18, 44, 132
 of industry, 2, 12, 80
 model of, 15
 open-ended, viii, x, 7, 9, 10, 12, 27, 38, 40, 48, 51, 52, 59, 77, 81, 89, 91, 114, 118
 universal, ix, 18
 world, vii, 2, 44
Human, vii, viii, x–xii, 2–8, 10–12, 21–25, 28, 31–33, 35–38, 40, 43, 45–48, 54–69, 71–77, 79–99, 101, 105, 109–111, 114, 116–128, 130, 132, 134–145
 inner life, 13, 14, 144, 145
 nature, 9, 17, 18, 20, 27, 54, 55, 67, 72, 76, 88, 96, 101n54
 outer life, 144, 145
Human senses, 71, 74–77, 79, 80, 93, 95, 144
Humanism, 11, 49, 51, 69–71, 74–77, 79, 85, 91, 92, 100, 106, 144

I

Idealism, 11, 51, 62, 64, 73–75, 79, 91
 objective, 21, 38, 45
 subjective, 41, 107
Incompleteness theorem, 17, 115, 128
 See also Godel; Mathematics
Industry, vii, 2, 70, 80, 84, 85, 89, 91, 118, 141
 See also Technology
Inexhaustible creativity, 111, 118, 124, 134, 135
 and open-ended history, 114
 and technological progress, 119–121

Infinite/unbounded/open-ended, 7, 9, 10, 27, 38, 40, 48, 51, 52, 59–61, 89, 91, 114, 118, 126–129, 134
  progress, viii, 2, 16, 46, 81; of science, viii, x, 2, 17, 115, 124, 130, 131, 138; of technology, ix, x, 2, 3, 11–13, 15, 77, 81, 95, 98, 120, 124, 127, 130
Infinity, viii, 93, 132, 133, 135
  actual, viii, 93, 132, 133, 135
  bad, 8, 12, 61
  in itself, 135
  and universality, 45
Intuition, 31, 32, 36–38, 38n41, 42, 128, 140
  intellectual, 7, 20, 30
  See also Productive imagination
Intuitionism
  and Brouwer, 32n31, 38n41, 139n76
  and Fichte, 38
  and mathematics, 38, 132
  and Weyl, 139n76
Invariants, 30, 93, 125–135
  of experience, viii, 135, 139
  of instrumental action, 105, 135
  of perception, 135
  relations, 135, 137
Isomorphism
  and invariants of experience, 93
  and nature, 93, 135
  and universality of man, 93

K
Knowledge, x, 4–6, 11n26, 19–48, 55, 103, 104, 106, 110, 115, 132, 135–145

L
Labor, vii, viii, x, 3, 4, 14, 22, 28, 45, 47, 49–77, 79–111, 118, 119, 121, 140, 142
  and capital, 80, 94
  emancipated, 80
  historical development of, 79
  ontology of, 12, 21, 26, 74–77, 95, 111
  technology-based, ix, xi, 10, 11, 76, 77, 80, 84, 109
Left-wing Hegelians, see Young Hegelians
Liberalism, 8, 13, 44, 47, 51, 61, 69
  political, x, xii, 10, 62
  post-Hegelian, 58
Lifeworld, 45, 46, 125, 140, 145
Limitations, 15–17, 25, 34, 119, 126, 128, 130
  epistemological, 120, 127
  physical, 120
Limits to growth, ix, xi, 3, 17n46, 81, 111
  scientific, 130
  technological, 121–124, 130
  See also Growth

M
Man, vii–viii, x, 2–6, 9–16, 18–21, 22n7, 23–27, 28n21, 35, 49, 51, 55, 56, 62, 64–69, 65n39, 71–77, 79–83, 85–87, 90–97, 99, 101, 104–107, 115, 128, 134–136, 138, 140, 140n80, 142, 143, 145
  as being for himself, 75, 81–83, 99, 105
  as material being, 26, 76, 77

as natural being, 11, 13, 67, 68, 75, 76, 82, 86, 91, 104
as social being, 67, 75
as universal being, vii, viii, 2, 3, 10–13, 13n33, 65, 65n39, 72, 74–76, 83, 91, 92, 96, 99, 101, 104
See also Being
Marxian circle, *see* Circle
Materialism, 11, 26, 41, 48, 51, 64, 67, 68, 70–72, 74, 79, 80, 84, 86, 89, 91, 92, 94, 98, 103–107, 109–111
contemplative, 11, 86, 104, 109
dialectical, 92, 98, 103–111
historical, 48, 51, 68, 71, 72, 75, 80, 89, 94, 98, 103, 106
naive, 110
Mathematics, xi, 3, 17, 25, 31–33, 31n30, 36, 113, 115, 124, 128, 129, 131–133, 136–139
and actual infinity, 132
endlessness of, 132
and formalism, 129
group, 32, 137
and intuitionism, 32n31, 38, 132
as transcendental psychology, 88, 129
as unbounded science, 129
and universality of, 133
Matter, 6, 7, 14, 16, 17, 21, 25, 31, 41, 43, 52, 53, 57, 65, 68, 75, 84, 91, 93, 97–99, 101, 103, 104, 106–110, 113, 115, 116, 118, 120, 126, 127, 131–133, 136, 137, 143n84
and dialectical materialism, 103
and Lenin, 104
and Marx, 26, 41, 68, 69, 76, 85, 91, 97–99, 104
and Plekhanov, 103

Matter and spirit, 7, 21, 41, 103
identity of, 7, 41
Metaphysics, 26–28, 35, 38, 43, 104, 105, 108, 114, 118, 140
Myth, 32, 41–46, 56, 58
and Christianity, 57
as a historical epiphenomenon, 59
and mythological consciousness, 46
and Schelling, 41–43, 45–47, 46n58, 57, 58
and Strauss, 46n58, 56–58

N
Natural necessity, 5, 26, 35
Nature, vii, viii, x, xi, 2–7, 9, 11–13, 17–22, 22n8, 25–28, 31–33, 35, 37, 38, 40–42, 45–47, 52–58, 64–69, 72, 75, 76, 79–96, 92n37, 101n54, 103–107, 104n62, 109–111, 109n76, 115, 116, 118–120, 123–126, 128–136, 138, 140, 140n80, 142–145
causality of, 20, 22, 22n8, 32, 40, 41, 45, 56
complexity of, viii, 128, 130, 131, 134–136
extended body of, 85, 92, 104, 138
human, 9, 17, 18, 20, 27, 54, 55, 67, 72, 76, 88, 96, 101n54
invariability laws of, 3, 25, 33
in itself, 5, 20, 32, 40, 86, 91, 92, 105
for man, 85, 92, 95, 106, 140
mastery of, 2–4, 25, 27, 47, 119, 136, 143
as mathematical manifold, 3, 25, 33
unity with, 86, 93, 143
Neo-Malthusians, 114, 121

172 SUBJECT INDEX

O

Objectification, Self-objectification, x, 11, 32, 68, 69n51, 75, 76, 79, 83, 87, 88, 93, 97, 98, 145
 and alienation, 93
 artistic, 145
 technological, 87, 145
Objective reality, 3, 33, 104, 108, 109, 114, 136
 and dialectical materialism, 103–111
 and Lenin, 104, 108
 and matter, 107, 109
 rationally organized, 114, 136
Ontology, viii, x, 7, 12, 13, 21, 26, 38n42, 40, 43, 51, 54, 74–78, 81, 85–87, 91, 92, 95, 104–106, 109, 111, 128, 134, 139, 140
 anthropocentric, 13, 26, 51, 81, 85–87, 92, 104, 111, 134
 of labor, 21, 26, 74–77, 95, 111
 and Marx, viii, 12, 21, 81, 87, 91, 92, 95, 104–106, 109, 111, 128, 134
 recapitulates technology, 12, 109
Optimism, viii, ix, 2, 3, 10, 16, 17, 28, 41, 44, 113
 historical, 3, 10, 28, 41, 44, 113
 technological, viii, ix, 2, 16, 17
Organ projection, xi, 32, 87, 88
 and Kapp, 32, 88
 and Marx, xi, 32, 87, 88
 and Noire, 32, 87, 88
 See also Kapp, Noire

P

Panlogism, 12, 21, 23, 43, 45
 Hegelian, 12, 21, 23, 45
Pantheism, 26, 38, 41, 52, 103
 and Spinoza, 26, 38, 41, 52, 103
 See also Spinoza

Philosophy, vii–xii, 2–4, 6–12, 6n15, 16, 19–21, 21n5, 25–28, 28n20, 31, 31n30, 32, 34–36, 38–45, 38n41, 46n58, 47–51, 53–64, 66–68, 73, 75, 79, 81, 86, 88, 91, 92, 98, 103–111, 114, 116–118, 128, 136, 138, 143, 144
 of art, 45
 European, 6, 20, 26, 28
 of history, viii, x–xii, 7, 9, 28, 38–41, 43, 44, 46n58, 47, 54, 56, 118
 liberal, x, 10
 modern, ix, xii, 2, 27
 of mythology, 45, 57
 of nature, 25, 26, 28, 40, 55
 of praxis, 68, 105–111, 114, 116–118
 of religion, 55
 of science, xi, xii
 of self-consciousness, 62
 of technology, xii, 3, 16, 32, 86n20
 transcendental, 11, 20, 31, 36, 41, 67, 73
Physics, 25, 33, 86, 108, 110, 113, 126–128, 130, 133, 137, 138
 mathematical, 33, 137
 newtonian, 25
 quantum, 108, 110, 133
Praxis, viii, x, 21, 34, 41, 45–48, 60, 62, 63, 68, 75, 92, 105–111, 114, 116–118
 historical, 45–48, 60, 63
 material, 63
 revolutionary, 21, 34, 41, 47
Private property, 71, 79, 89, 94
 essence of, 79
 subjective and objective aspects of, 80
Productive imagination, 32, 37, 43, 46, 88, 139, 145

Progress, vii–xi, 2–17, 20, 21, 27, 28,
    40, 46, 47, 51, 56, 58, 60, 61,
    70, 77, 80, 81, 86–90, 95, 98,
    99, 114–116, 119–121, 123,
    124, 127–131, 134, 136, 138
  directional, vii, x, 2, 3, 16, 58
  historical, 5, 6, 8, 10, 12, 20, 21,
    27, 28, 35, 40, 51, 56, 60
  infinite, viii, 2, 4, 11–13,
    16, 46, 124
  qualitative, 130
  scientific (of science), xi, 17, 27,
    114, 115, 127, 129–131, 134,
    136, 138; endlessness of, 132,
    134; limits of (to), 129
  technological, vii–xi, 2, 3, 14–17,
    77, 80, 81, 86–88, 98, 115,
    121, 123, 124, 127, 128, 130
  unlimited, 124
  See also Science; Technology
Proletariat, 13, 64, 70, 71, 74, 80,
    94–96, 98–101, 99n48
  dictatorship of, 98, 99, 101
  as a historical universal, 99
  psychological emancipation of, 95
  as a universal class, 74, 95, 96, 99

R
Rationalism, ix, xii, 3, 4, 7, 20, 22,
    22n8, 25–27, 29, 34, 38, 61, 97,
    114, 135, 136, 140, 142
  and Bachelard, 135
  Cartesian, ix, 3, 4, 22, 27, 29, 38,
    136, 140
  European, ix, xii, 7, 26, 142
  and invariants of experience, 135
  and Kant, 26, 29, 136
  and relativism, 3
Realm of freedom, vii, 64, 68, 71, 72,
    80n3, 90, 91, 114

  and Kant, 68
  and Marx, 64, 114
  and realm of necessity, 72, 90
Realm of necessity, 4, 12, 34,
    40, 68, 90
Reason, xii, 2–5, 10, 17, 19–22, 21n5,
    24, 25n13, 26–36, 39–42, 45–48,
    50–53, 55, 56, 59–66, 85, 93,
    105–107, 113–145
  critique of, 28, 46, 52, 115
  invariability laws of, 33
  practical, 5, 20, 34–36, 56
  pure, 20, 32, 34, 35, 39, 42, 46, 53,
    85, 105, 107
  theoretical, 5, 34, 35, 47, 119,
    120, 137

S
Science, viii–xii, 2–4, 17, 17n46,
    19–21, 25, 27, 31–33, 40, 46,
    52, 61, 62, 65, 68, 81, 93, 98,
    111, 113–118, 114n1, 119n14,
    120–138, 120n19, 137n70,
    142, 143
  bounded; epistemologically, 127;
    ontologically, 126, 127
  end of, ix, xi, 3, 17, 17n46, 81,
    111, 124–135
  of knowledge, 21, 36, 43 (*see
    also* Fichte)
  open-endedness of, 134
  a priori, 33
  saturation of, 115, 125
  unbounded; epistemologically, 3,
    126; ontologically, 126;
    universality of, 93, 133
  universal, 3, 20, 21, 25, 65, 68,
    116, 117
Scientific communism, viii, 7, 70, 74,
    80, 81, 100

Scientific progress, xi, 17, 27, 114, 115, 127, 129, 131, 134, 136, 138
　endlessness of, 134
　limits of (to), xi, xii, 17, 98, 115, 129, 129n45
　See also Progress
Second law of thermodynamics, 111, 114, 115, 118–120, 124
　and closed system, 119, 120, 124
　and entropy, 119, 120, 124
　and open system, 120, 124
　and selection principle, 119
　tektological interpretation of, 111, 114
　See also Entropy
Selection, 118–120, 118n14
　of the elements of experience, 119
　negative, 119
　positive, 119, 120
　principle, 118, 118n14, 119
Self-externalization, viii, 12, 12n30, 68, 76, 81, 84–87, 89–98, 144
　and objectification, 75, 87
　and self-reference, 75, 76
　technological, viii, 12, 85, 86, 144
Self-positing, 24n13, 139, 140, 145
　and ego, 24n13
　and history, 24n13, 139, 140, 145
　and Schelling, 140
Self-reference, x, 11, 24, 75, 76, 82, 83, 98, 136
　and dialectic, x, 24, 98
　and Marx, 76, 82, 83
　and objectification, 11, 75, 76, 98
　and technological growth, 81
Self-reproducing automata, 84
Sense of having, viii, 27, 70, 71, 79, 95, 96, 98, 144
　See also Human senses; Marx
Social anthropology, 48–77, 105
Social humanity, vii, 13, 70, 71
　vs. civil society, 70, 71

　vs. and Hess, 71
　vs. and Marx, 71
Socialism, x, 13, 48, 51, 68–74, 90, 93, 95, 100, 102
　evolutionary, 13
　scientific, 100
Social yoga, 144
Society, 2, 5, 8, 9, 12–15, 27, 43, 47, 51, 64, 67, 69–71, 74, 80, 85, 90, 91, 94, 95, 99, 101, 109, 116, 117, 143, 144
　capitalist, 64, 71, 80, 109
　civil, 5, 8, 9, 13, 15, 43, 47, 70
　closed, 144
　communist, 91, 95
　humane, 51, 64, 69–71, 74, 99, 101
　open, 144
　socialist, 90, 94, 99
Software (design), 141, 142
　commercial, 142
　consumer technology, 141
　free, 142
　open-source, 142
Species, 5, 9, 35, 56, 57, 57n14, 59, 63–67, 65n39, 69, 71–74, 74n63, 79, 81–84, 88, 91, 94, 99, 145
　being, 11, 64, 65, 65n39, 72, 74, 81–83, 94, 99
　existence, 9, 65, 71, 91
　life, 64, 74, 83, 143, 145
Spirit, x, xi, 4–8, 10, 11, 13, 16, 19–21, 24n13, 26, 28, 31n30, 35, 36, 38, 40, 41, 43, 44, 47, 49, 50, 52, 53, 55, 57, 60–62, 66, 67, 94, 96, 97, 100–103, 105, 108, 109, 114, 121, 124, 127, 136, 137, 144
　and matter, 21, 41, 43, 57, 103, 136
Standing reserve, 86, 140, 141, 141n81, 145
　attitude of, 141

culture of, 141
and enframing, 86
and technology, 86, 140
world of, 145
See also Heidegger
State, vii, 8–10, 14, 29, 30, 33, 37, 38, 41–45, 49, 53, 55, 59, 61, 96, 97, 100–103, 101n53, 115, 118, 119, 122, 123, 126, 127, 133, 134
and church, 50, 101
and civil society, 8, 9, 43
as concrete universal, 21, 43
as ethical idea, 43, 50
and Hegel, 7–9, 21, 45, 50, 55, 61, 97
and liberalism, 8, 61
as spirit, 8, 43, 55, 61, 102
and statism, 8, 101
Subjectivism, 41, 138
Substance, xiii, 3, 20, 22, 22n8, 23, 25, 26, 30, 31, 41, 42, 48, 52, 77, 85, 88, 96, 103–106, 109–111, 134
material, 20, 25, 26, 110, 111
thinking, viii, 3, 20, 22n8, 23, 25, 26, 30, 31, 52, 77, 96, 111
System, viii, x, 6, 9, 11, 15, 16, 21, 26, 34, 38, 39, 41, 43–45, 47, 49–57, 59, 60, 64, 66, 74, 83, 88, 95–97, 99–101, 99n48, 105, 107, 109, 114, 115n2, 116–120, 122–124, 122n22, 126, 127, 131, 133, 135
closed, 119, 120, 124, 135
epistemologically open, 120, 124

T
Technological control, *see* Control
Technological determinism, 12, 141
Technological growth, *see* Growth

Technology, vii–x, xii, 2, 11–13, 15, 16, 27, 32, 47, 76, 80–89, 91n31, 98, 109, 118, 121–125, 127, 128, 130, 136, 140–143, 142n82
evolution of, xi, 87
and information technology (IT), 141
limits of (technological growth), xi, xii, 79–111, 114, 115, 121–125
and science, viii, x, xii, 2, 27, 98, 121–123, 130
and technoscience, 17, 125, 137, 138
See also Industry
Technoscience, viii, 17, 93, 124, 125, 125n34, 137, 138
Tektology, xi, 111, 114, 116, 117
and cybernetics, xi, 84n14, 114, 116
and dialectic, 116, 117
and thermodynamics, 111, 114
See also Bogdanov; General theory of organization
Theology, ix, 3–4, 9, 19, 43, 48, 50–57, 64, 66, 69, 106
esoteric, 43, 48, 53, 64, 106
and Hegel, 9, 48, 54, 57, 64, 66, 106
and Kant, 53, 55–57
rational (reconstruction of), 52, 53, 55–57
and Spinoza, 50, 52, 53, 103
and Strauss, 50, 56, 57
and theological exegesis, 47, 50, 55, 64
traditional, 53, 57
Time, ix, xi, 3, 11, 13–17, 19, 21, 25, 26, 32, 36, 37, 38n41, 39–42, 44–46, 50, 56, 60–62, 66, 69, 79, 81, 87–90, 94, 95, 97, 100, 100n55, 113, 114, 120, 120n18,

122, 126, 131, 133, 133n56,
    138–140, 142, 145
  and Brouwer, 38n41, 139
  and Fichte, 39
  historical, 46, 101n55, 145
  and mathematics, 25, 32, 133
  as *an a priori form of sensibility*, 32
Tradition, ix, 7, 8, 19, 21, 23, 25, 43,
    45, 46, 49, 52, 73, 74, 85, 107,
    123, 132, 140–143
  Aristotelian, 19
  and culture, ix, 141, 145
  Eastern, 142
  Enlightenment, 49, 73
  Platonic, 19
  transcendental, 21, 43, 46, 85, 140
Transcendental methodology/
    transcendentalism, 6, 28, 33, 34,
    38, 40, 41, 43, 67, 104, 107,
    135n63, 139
  Brouwer's, 139
  Fichte(an), 34, 38, 40, 41
  Kant(ian), 6, 28, 33, 67, 104,
    107, 135n63

U

Uncertainty principle, 33, 110, 137
  and Bohr, 33
  and complementarity, 33
  and Heisenberg, 33, 110, 137
Universality, universal, vii–xi, 1–5,
    8–10, 12, 13, 13n33, 15, 17, 18,
    20–22, 22n8, 25–27, 31, 32, 43,
    44, 47, 50, 51, 63–68, 71–76,
    80–83, 84n14, 86, 88, 89,
    91–96, 99–102, 104, 105, 111,
    115–117, 133,
    135–137, 142–144
  human, 10, 72, 81, 93, 137, 142
  of labor, 91, 93

of man, viii, 4, 9, 13, 22n8, 51, 72,
    74, 75, 83, 86, 93, 105,
    115, 135
  and Marx, viii, 10, 22, 26, 64, 72,
    74–76, 81, 83, 86, 88, 89,
    96, 99, 104
  of moral law, 5
  and objectivity, 33, 63, 82, 86, 105
  of science, 3, 20, 25, 65, 68, 93,
    116, 117

W

Will, 5, 61
  arbitrary, 35, 43, 50
  free, 5, 61
  general, 8, 8n21, 21, 50
  good, 18, 35, 96
  universal, 18, 50

Y

Young Hegelians, 34, 41, 63
  Bauer, Bruno, x, 9–11, 47, 48, 51,
    58–64, 61n24, 63n29, 68, 70
  Engels, Friedrich, 10, 12, 51, 59,
    62, 63, 70, 71, 73n61, 75, 85,
    89, 92n37, 94, 102–104,
    104n62, 106, 107, 121
  Hess, Moses, x, 48, 51, 61–63,
    68–71, 69n47, 109n75
  Marx, Karl, vii–xii, 1n1, 3, 4, 7,
    10–12, 10n25, 11n26, 21,
    21n5, 22, 26–28, 31, 32, 34,
    35, 41, 45–49, 47n59, 51, 58,
    58n18, 59, 61–64, 63n29,
    67–77, 68n45, 69n47, 74n63,
    79–111, 92n37, 104n62,
    108n71, 109n75, 114,
    116–121, 128, 134, 138, 141,
    142, 144

Ruge, Arnold, x, 51, 61–63
Stirner, Max, 9, 72–74, 73n61
Strauss, David, x, 46n58, 48, 50, 54–59, 64, 65n39

Z

Zero growth, 16, 95, 115, 122, 123
   and Club of Rome, 16, 114, 123
   and technology, 115, 123
   See also Growth